JAAVSO

The Journal of
The American Association
of Variable Star Observers

Volume 43
Number 1
2015

AAVSO
49 Bay State Road
Cambridge, MA 02138
USA

ISSN 0271-9053

Publication Schedule

The Journal of the American Association of Variable Star Observers is published twice a year, June 15 (Number 1 of the volume) and December 15 (Number 2 of the volume). The submission window for inclusion in the next issue of JAAVSO closes six weeks before the publication date. A manuscript will be added to the table of contents for an issue when it has been fully accepted for publication upon successful completion of the referee process; these articles will be available online prior to the publication date. An author may not specify in which issue of JAAVSO a manuscript is to be published; accepted manuscripts will be published in the next available issue, except under extraordinary circumstances.

Page Charges

Page charges are waived for Members of the AAVSO. Publication of unsolicited manuscripts in JAAVSO requires a page charge of US $100/page for the final printed manuscript. Page charge waivers may be provided under certain circumstances.

Publication in *JAAVSO*

With the exception of abstracts of papers presented at AAVSO meetings, papers submitted to JAAVSO are peer-reviewed by individuals knowledgable about the topic being discussed. We cannot guarantee that all submissions to JAAVSO will be published, but we encourage authors of all experience levels and in all fields related to variable star astronomy and the AAVSO to submit manuscripts. We especially encourage students and other mentees of researchers affiliated with the AAVSO to submit results of their completed research.

Subscriptions

Institutions and Libraries may subscribe to JAAVSO as part of the Complete Publications Package or as an individual subscription. Individuals may purchase printed copies of recent JAAVSO issues via Createspace. Paper copies of JAAVSO issues prior to volume 36 are available in limited quantities directly from AAVSO Headquarters; please contact the AAVSO for available issues.

Instructions for Submissions

The *Journal of the AAVSO* welcomes papers from all persons concerned with the study of variable stars and topics specifically related to variability. All manuscripts should be written in a style designed to provide clear expositions of the topic. Contributors are encouraged to submit digitized text in MS WORD, LATEX+POSTSCRIPT, or plain-text format. Manuscripts may be mailed electronically to journal@aavso.org or submitted by postal mail to JAAVSO, 49 Bay State Road, Cambridge, MA 02138, USA.

Manuscripts must be submitted according to the following guidelines, or they will be returned to the author for correction:

- Manuscripts must be:
    1) original, unpublished material;
    2) written in English;
    3) accompanied by an abstract of no more than 100 words.
    4) not more than 2,500–3,000 words in length (10–12 pages double-spaced).

- Figures for publication must:
    1) be camera-ready or in a high-contrast, high-resolution, standard digitized image format;
    2) have all coordinates labeled with division marks on all four sides;
    3) be accompanied by a caption that clearly explains all symbols and significance, so that the reader can understand the figure without reference to the text.

Maximum published figure space is 4.5" by 7". When submitting original figures, be sure to allow for reduction in size by making all symbols, letters, and division marks sufficiently large.

Photographs and halftone images will be considered for publication if they directly illustrate the text.

- Tables should be:
    1) provided separate from the main body of the text;
    2) numbered sequentially and referred to by Arabic number in the text, e.g., Table 1.

- References:
    1) References should relate directly to the text.
    2) References should be keyed into the text with the author's last name and the year of publication, e.g., (Smith 1974; Jones 1974) or Smith (1974) and Jones (1974).
    3) In the case of three or more joint authors, the text reference should be written as follows: (Smith et al. 1976).
    4) All references must be listed at the end of the text in alphabetical order by the author's last name and the year of publication, according to the following format: Brown, J., and Green, E. B. 1974, *Astrophys. J.*, **200**, 765. Thomas, K. 1982, *Phys. Rep.*, **33**, 96.
    5) Abbreviations used in references should be based on recent issues of the *Journal* or the listing provided at the beginning of *Astronomy and Astrophysics Abstracts* (Springer-Verlag).

- Miscellaneous:
    1) Equations should be written on a separate line and given a sequential Arabic number in parentheses near the right-hand margin. Equations should be referred to in the text as, e.g., equation (1).
    2) Magnitude will be assumed to be visual unless otherwise specified.
    3) Manuscripts may be submitted to referees for review without obligation of publication.

Online Access

Articles published in JAAVSO, and information for authors and referees may be found online at: http://www.aavso.org/journal-aavso

© 2015 The American Association of Variable Star Observers. All rights reserved.

The Journal of the American Association of Variable Star Observers
Volume 43, Number 1, 2015

Editorial

The Rise and Fall and Rise of the David Dunlap Observatory
John R. Percy — 1

Variable star analysis

Long Term Photometric and Spectroscopic Monitoring of Semiregular Variable Stars
Robert R. Cadmus, Jr. — 3

Photometric Analyses and Spectral Identification of the Early-Spectral Type W UMa Contact Binary V444 Andromedae
Ronald G. Samec, Russell Robb, Danny R. Faulkner, Walter Van Hamme — 8

The Curious Case of ASAS J174600-2321.3: an Eclipsing Symbiotic Nova in Outburst?
Stefan Hümmerich, Sebastián Otero, Patrick Tisserand, Klaus Bernhard — 14

New Variable Stars Discovered by the APACHE Survey. II. Results After the Second Observing Season
Mario Damasso, Lorenzo Gioannini, Andrea Bernagozzi, Enzo Bertolini, Paolo Calcidese, Albino Carbognani, Davide Cenadelli, Jean Marc Christille, Paolo Giacobbe, Luciano Lanteri, Mario G. Lattanzi, Richard Smart, Allesandro Sozzetti — 25

UXOR Hunting among Algol Variables
Michael Poxon — 35

Sudden Period Change and Dimming of the Eclipsing Binary V752 Centauri
Anthony Mallama, Hristo Pavlov — 38

The δ Scuti Pulsation Periods in KIC 5197256
Garrison Turner, John Holaday — 40

Early-Time Flux Measurements of SN 2014J Obtained with Small Robotic Telescopes: Extending the AAVSO Light Curve
Björn Poppe, Thorsten Plaggenborg, WeiKang Zheng, Isaac Shivvers, Koichi Itagaki, Alexei V. Filippenko, Jutta Kunz — 43

Recently Determined Light Elements for the δ Scuti Star ZZ Microscopii
Roy Andrew Axelsen, Tim Napier-Munn — 50

A Photometric Study of ASAS J184708-3340.2: an Eclipsing Binary with Total Eclipses
Robert C. Berrington, Erin M. Tuhey — 54

A Binary Model for the Emission Line Star FX Velorum
Mel Blake, Maisey Hunter — 59

Comparison Between Synthetic and Photometric Magnitudes for the Sloan Standard Stars
Anthony Mallama — 64

Table of Contents continued on following pages

Variable star data

Revised Light Elements of 78 Southern Eclipsing Binary Systems
Margaret Streamer, Jeff Byron, David J. W. Moriarty, Tom Richards, Bill Allen, Roy Axelsen, Col Bembrick, Mark Blackford, Terry Bohlsen, David Herald, Roland Idaczyk, Stephen Kerr, Ranald McIntosh, Yenal Ogmen, Jonathan Powles, Peter Starr, George Stockham — 67

Recent Maxima of 67 Short Period Pulsating Stars
Gerard Samolyk — 74

Recent Minima of 149 Eclipsing Binary Stars
Gerard Samolyk — 77

Instruments, methods, and techniques

Video Technique for Observing Eclipsing Binary Stars
Hristo Pavlov, Anthony Mallama — 80

History and biography

Margaret Harwood and the Maria Mitchell Observatory
James W. Hanner — 84

Some Personalities from Variable Star History
Edited by Thomas R. Williams, Michael Saladyga — 86

Some Personal Thoughts on TV Corvi
David H. Levy — 102

Abstracts of Papers and Posters Presented at the 103rd Annual Meeting of the AAVSO, November 6–8, 2014, Woburn, Massachusetts

General Paper Session Part I

Betelgeuse Period Analysis Using VSTAR
Frank Dempsey — 105

EE Cep Winks in Full Color
Gary Walker — 105

Transient Pulsation of Sirius
Kangujam Yugindro Singh, Irom Ablu Meitei — 105

η Carinae Continues to Evolve
John C. Martin — 105

The Trend in the Observation of Legacy Long Period Variable Stars (poster)
Robert Dudley — 105

Analysis of Hα lines in ε Aurigae post-eclipse (poster)
Shelby Jarrett, Cybil Foster — 106

Table of Contents continued on following pages

Discovery of Five Previously Misidentified BY Draconis Stars in ASAS Data (poster)   106
Jessica Johnson, Kristine Larsen

AAVSO and the International Year of Light (poster)   106
Kristine Larsen

Precision Photometry of Long Period Variable Stars: Flares and Bumps in the Night (poster)   106
Dale Mais

Transformation: Adjusting Your Data to the Standard Photometric Framework (poster)   107
George Silvis

The Eggen Card Project (poster)   107
George Silvis

Visual Spectroscopy of R Scuti (poster)   107
Lucian Undreiu, Andrew Chapman

General Paper Session Part II

Parallel Group and Sunspot Counts from SDO/HMI and AAVSO Visual Observers   107
Rodney Howe, Jan Alvestad

Going Over to the Dark Side   108
David Cowall

Photometry Transforms Generation with PTGP   108
Gordon Myers, Ken Menzies, George Silvis, Barbara Harris

Using VPHOT and PTGP to Generate Transformation Coefficients   108
Ken Menzies, Gordon Myers

General Paper Session Part III

Observational Activities at Manipur University, India   109
Kangujam Yugindro Singh, Irom Ablu Meitei, Salam Ajitkumar Singh, Rajkumar Basanta Singh

A Report on West Mountain Observatory Observations for the KELT Follow-up Observing Network   109
Michael Joner

Visual Observing: New Ideas for an Old Art?   109
David Turner

America's First Variable Star   109
John Toone

General Paper Session Part IV

The Future of Visual Observations in Variable Star Research: 2015 and Beyond   109
Mike Simonsen

The Life of Albert Jones   109
John Toone

Table of Contents continued on next page

Special Paper Session Part I

Why Do Some Cataclysmic Variables Turn Off?
Kent Honeycutt — 110

Special Paper Session Part II

Before the Giants: APASS Support to Ambitious Ground-based Galaxy Investigations and Space Missions Serching for Exo-Earths
Ulisse Munari — 110

APASS and Galactic Structure
Stephen Levine — 110

Astronomical Photometry and the Legacy of Arne Henden
Michael Joner — 110

Special Paper Session Part III

A Journey through CCD Astronomical Imaging Time
Richard Berry — 110

Collaborations with Arne on Cataclysmic Variables
Paula Szkody — 110

The History of AAVSO Charts, Part III: The Henden Era
Mike Simonsen — 111

Arne's Decade
Gary Walker — 111

Editorial

The Rise and Fall and Rise of the David Dunlap Observatory

John R. Percy
Editor-in-Chief, *Journal of the AAVSO*

Department of Astronomy and Astrophysics, University of Toronto, Toronto, ON M5S 3H4, Canada; john.percy@utoronto.ca

Received June 4, 2015

As I write this editorial, I am also preparing a presentation for a one-day symposium on June 13 marking the 80th anniversary of the founding of the David Dunlap Observatory (DDO; http://theddo.ca). Its story has many connections with and lessons for the AAVSO.

DDO was the dream of Professor Clarence Augustus Chant (1865–1956). Chant graduated from the University of Toronto in physics in 1890. In 1892, he rejoined the Department of Physics as a lecturer. Over the next four decades, he introduced courses in astrophysics (previously, only courses in mathematical and practical astronomy had been offered in the university), established a degree stream in physics and astronomy, and eventually a separate Department of Astronomy. He was also a guiding light for the mostly-amateur Royal Astronomical Society of Canada (RASC; http://rasc.ca) for half a century. It's not surprising that University of Toronto astronomers have enjoyed a respectful and productive partnership with amateur astronomers ever since, through the RASC, the AAVSO, and other organizations. Chant was also a prolific and successful communicator and popularizer of astronomy. His book *Our Wonderful Universe* (1928) was a best-seller in several languages.

Throughout his career, Chant dreamed of establishing a major observatory for the university and for the city. In 1921, after one of his many public lectures, lawyer and mining executive David Dunlap approached him and expressed an interest in astronomy and in the observatory project. Sadly, Dunlap died before any further discussions took place. In 1926, Chant got up the courage to approach his widow, Jessie Donalda Dunlap, and asked if she might consider donating an observatory to the university as a memorial to her late husband. She agreed and, on May 31, 1935, DDO opened, housing a 1.88-m (74-inch) reflector, the second-largest telescope in the world at the time. This story illustrates one of the many potential benefits of astronomy outreach and communication! More obvious reasons to do outreach are: to increase public awareness and understanding of astronomy; to increase public science literacy in general; to present a positive image of astronomy and astronomers; to justify public spending on astronomy; to attract young people (and amateurs) to astronomy; or simply because outreach is fun!

Initially, the DDO's research was in the area of stellar radial velocities, but the study of binary and variable stars soon became central to its work. Helen Sawyer Hogg (AAVSO president 1939–1941) was world-renowned for her research, catalogues, and bibliographies of variable stars in globular clusters, work which has been continued by her former student Christine Clement. Smaller telescopes, equipped for photometry, were added. These, together with the 1.88-m spectroscopic "Great" telescope, led to long-term studies of Mira, RV Tauri, R CrB, RS CVn, Be and shell, peculiar A, and especially Population I and II Cepheid variables. Many of these studies were carried out as thesis projects by graduate students who subsequently became leaders in variable star research. Most have been "friends" of the AAVSO in one way or another. DDO director (1978–1988) Don Fernie was internationally known for his research on the Cepheid period-luminosity relation, now known as the Leavitt Law. Radio astronomer and DDO director (1988–1999) Ernie Seaquist studied radio emission from a variety of unusual variable and binary stars such as novae and symbiotic stars. Slavek Rucinski was the "father" of the MOST (http://en.wikipedia.org/wiki/MOST (satellite)) and BRITE (http://brite-constellation.at) variable star space telescopes. The most famous advance at DDO was Tom Bolton's identification of the optical counterpart of X-ray source Cyg-X1 as the first black hole binary.

DDO's success in these fields was due to its availability for sustained, systematic surveys and studies of individual stars, using both spectroscopic and photometric techniques. Longterm spectroscopic studies of variable stars are still rare. It helps to be a "local" observatory, with time allocation under the control of a single institution or organization. Of course, "sustained and systematic" is the key to the importance of the AAVSO's International Database of observations.

By the 1990s, however, stellar astronomy fell out of style, at least in North America—despite the importance of the sun and stars to all areas of science. Weather and light pollution made DDO a less-than-ideal site. University of Toronto astronomers and their students were observing from Hawaii, Chile, or from space. DDO was used less and less for research, though public programs continued, with the help of the RASC's Toronto Centre (http://www.rascto.ca). Its demise appeared to be imminent. Various groups, with different agendas, campaigned to save the observatory but, in 2008, with the agreement of the Dunlap family, the university sold the observatory to Metrus Development Inc.

If the story ended there, it would be a sad one. In 2009, however, the university invested the proceeds of the DDO sale—about $60,000,000 Canadian—in a new Dunlap Institute (http://www.dunlap.utoronto.ca), which has carried forward the Dunlap name and bequest into the 21st century in a very effective way. Donor interest and recognition is a key element of philanthropy, and we are delighted that the Dunlap family has maintained a strong interest in astronomy, and in the long-term impact of their bequest. The Dunlap Institute's mission is (1) to develop innovative instrumentation, including for the world's largest telescopes, (2) to help train the next generation of astronomers, and (3) to foster public interest and engagement in science. In partnership with the Department of Astronomy and Astrophysics, it is systematically building up a group of bright young faculty who, with Dunlap postdocs and graduate students, work at the frontiers of astronomy.

At the same time, the RASC Toronto Centre was successful in convincing Metrus to give it the use of the observatory to continue its public education and outreach programs. Every Saturday evening between May and October there are two public tours. They include a short non-technical talk by people such as me, followed by a tour of and look through the telescope, supplemented by exhibits and displays and state-of-the-art audiovisuals. Smaller telescopes are set up on the lawn. There are programs for schools, for Scouts and Guides, family nights with a "Space Crafts" room, and programs for special events such as meteor showers, the 2012 Transit of Venus, and Astronomy Day. In 2015, the dozens of "DDO volunteers" received the RASC's national Qilak Award for excellence in astronomy outreach and communication (http://www.rasc.ca/sites/default/files/2015_QilakDDO_volunteers.pdf). Also in 2015, Metrus officially donated the observatory to the RASC Toronto Centre—for better or worse!

There must be many "orphan" small-to-medium-sized telescopes in accessible locations across North America and beyond, and many astronomical organizations and clubs looking for a challenging outreach project (and maybe a telescope to do variable star research with). At the same time, there are millions of people, young and old, who could be inspired by looking through a telescope. The work of the RASC Toronto Centre is an excellent example of what can be done. I hope it inspires other astronomy groups to do likewise.

Long Term Photometric and Spectroscopic Monitoring of Semiregular Variable Stars

Robert R. Cadmus, Jr.
Department of Physics, Grinnell College, Grinnell, IA 50112; cadmus@grinnell.edu

Received September 12, 2014; revised December 8, 2014; accepted December 9, 2014

Abstract The understanding of semiregular variable stars presents a number of challenges that can be addressed by consistent long term photometric and spectroscopic monitoring. The observing program at Grinnell College has generated a large body of such data that has been used to investigate modes of pulsation, the role of dust, the possible role of chaos, and other issues. This paper summarizes these efforts and encourages other observers to help maintain the continuity of these data sets.

1. Introduction

The semiregular variable stars are pulsating red giants. Although a great deal of observational and theoretical effort has been expended in understanding these stars, and substantial progress has been made, a number of significant questions remain. There is clearly a great deal left to do and a great deal left to learn. It is likely that observations over extended periods of time will play an important role in our attempt to better understand these perplexing stars; the observing program at Grinnell College was designed to contribute to that effort.

The purpose of this paper is not to address all the scientific issues that are involved in this research program or to present final conclusions, but rather to make others aware of the work that we have been doing and the data that we have acquired, and to solicit the contribution of additional photometry from others in the future.

2. Background

The semiregular variable stars in this program are red giants that lie on the asymptotic giant branch (AGB) of the Hertzsprung-Russell diagram. They have extended atmospheres in which a variety of physical processes result in variations of luminosity and spectrum typically on timescales of hundreds of days, although much slower variations sometimes occur. The semiregular variables are similar to the Mira variables and have traditionally been distinguished from them on the basis of smaller visual light amplitudes and less consistent light curve shapes, but the exact nature of the relationship is still a matter for investigation. The study of these stars has been complicated both by the complexity of the light curves and by the array of physical processes that are probably at work. Much of the research on semiregular variables has focused on several related questions. (1) What is the nature and physical cause of the light variations? (2) What is the significance of the multiple sequences that are observed in the period-luminosity relations for long period variables? (3) What is the origin of the "long secondary periods"?

As the name suggests, the light variations of the semiregular variables are somewhat, but not entirely, regular. In fact, the light curves can be very messy. The dominant variation is periodic and additional periodicities are sometimes conspicuous. The application of Fourier analysis and other techniques in numerous studies (see, for example, Mattei *et al.* 1997; Kiss *et al.* 1999; Percy *et al.* 2003; Percy and Tan 2013) has shown that many of these stars are truly multiperiodic and the presence of periodicity suggests that systematic physical processes are at work. The exact nature of those processes is still a matter of some debate, but the prevailing interpretation of the primary light variations in semiregular variables is that these stars are radially pulsating, as is believed to be the case for some other classes of variable stars. Radial velocity measurements (Cummings *et al.* 1999; Lebzelter *et al.* 2000) strongly suggest that this is the case, as do the observations of variations in radius (Perrin *et al.* 1999) and theoretical models (Xiong and Deng 2007; Ireland *et al.* 2011). Furthermore, the spectra of these stars sometimes include hydrogen emission lines that appear to be the result of pulsation-induced shocks (Wood 1974; Willson 1976). After an extended debate (Wood 1995; Percy and Polano 1998; Willson and Marengo 2012) a consensus emerged that the dominant mode of pulsation is the fundamental for the Miras and an overtone for the semiregulars. Early work at Grinnell contributed to this conclusion (Cadmus *et al.* 1991). More recent observational and theoretical work has refined this picture but the complexity of the light variations and other characteristics of these stars leave a number of questions unresloved. For example, the association of modes with the Mira and semiregular classifications seems not to be absolute (Soszyński *et al.* 2013a), there are variations in the amplitudes of the light curves (Kiss *et al.* 2000; Percy and Abachi 2013), the origin of the long secondary periods is still unclear (Wood *et al.* 2004), and the degree to which the shorter-period variations are associated with overtone pulsation is still an interesting question.

Theorists have made a noble effort to include many of the complex physical processes that are simultaneously in play in a radial pulsation model, but a complete description is not yet in hand. In addition, the presence of several processes in addition to those directly related to radial pulsation has been suggested and debated, often in the context of the vexing long secondary periods discussed below. These include non-radial pulsation (Stello *et al.* 2014), convection (Stothers 2010), rotation (Wood *et al.* 2004, which also discusses other effects), binarity (Nicholls *et al.* 2009), planets (Berlioz-Arthaud 2003), and chaos (Buchler *et al.* 2004). There is clearly much work to be done on these stars by both observers and theorists.

The existence of a period-luminosity (P-L) relation for long period variables was established long ago (Gerasimovič 1928) but the availability of large surveys such as OGLE and MACHO in the 1990s revolutionized this area of research by generating vast archives of light curves. These observations of stars in the Large Magellanic Cloud have been especially valuable because those stars are all at approximately the same distance so their relative luminosities can be determined with good accuracy. These new data led to the discovery of multiple P-L relations, called sequences, for various groups of stars, including the semiregulars (Wood et al. 1999). The Hipparcos mission, which generated greatly improved parallaxes for many stars, led to improvements in the P-L relations for galactic stars and showed that the same sort of sequences are present in that population (Soszyński et al. 2013b). A significant challenge has been to identify with certainty what distinguishes the stars that make up each of the handful of sequences. There is evidence that some of the sequences correspond to stars pulsating in different modes and others correspond to other phenomena (Wood 2000; Soszyński et al. 2007), but the distinctions are not always as sharp as one might wish. For example, the semiregulars appear to populate two of the sequences (Soszyński et al. 2013a).

One of the sequences in the P-L diagram is associated with the long secondary periods—light variations on a time scale that is roughly ten times longer than the likely pulsational periods (Wood et al. 1999). These variations in some semiregulars have been known for a long time (Payne-Gaposchkin 1954) and have been the subject of much research in recent years. A number of explanations for this phenomenon have been offered (Nicholls et al. 2009; Wood et al. 2004) but none has gained widespread acceptance.

While much progress has been made, understanding these aspects of the nature of the semiregular variables has been challenging, leaving significant questions only partially answered and offering ample opportunities for further research.

3. Observations

Grinnell College's Grant O. Gale Observatory began operation in 1983 and provided the opportunity to create a new astronomical research program. With the advice and encouragement of Lee Anne Willson of Iowa State University and Janet Mattei of the AAVSO we began a project to help resolve the question of the modes of pulsation of the Mira and semiregular stars. The AAVSO database showed that some semiregular variables experienced quiescent episodes during which the amplitude of the brightness variations dropped into the noise in the AAVSO visual observations. The scientific question was whether these events might be cases of mode switching, in which the quiescent episodes are characterized by low-amplitude variations with periods different from those of the normal oscillations. If so, precise observations of those low-amplitude variations might lead to mode identifications through comparison of observational period ratios with those predicted by theoretical models. Some of our results support this idea, but over the years the picture has become more complex.

The 39 stars in our program (a few of which have been observed over less than the full duration of the project) are listed in Table 1. Nine of the stars are carbon stars (EU And, S Aur, U Cam, WZ Cas, RS Cyg, TT Cyg, V778 Cyg, RY Dra, and UX Dra). Because the quiescent episodes do not occur predictably, more than a few stars had to be monitored to have a chance of catching a reasonable number of quiescent episodes in a realistic length of time.

The photometric observations were made with the 0.61-m telescope at Grinnell College's Grant O. Gale Observatory using a photoelectric photometer incorporating an uncooled 1P21 photomultiplier. This instrument has been used throughout the project to maintain consistency. V- and B-band data were acquired differentially relative to the comparison and check stars in Table 1. No significant variations of the check stars relative to the comparison stars were observed, although in a few cases there may be hints of variability at a very low level that does not compromise the program star light curves. The data were reduced by subtracting the sky background, correcting for both first and second order atmospheric extinction, and transforming to the Johnson UBV system. An effort was made to avoid major gaps in the data by observing in twilight when necessary and applying a nonlinear correction for the temporal variation in sky

Table 1. Program and Comparison Stars.

Program Star	Comparison Star	Check Star
RU And	TYC 2814-1513-1	TYC 2814-0038-1
RV And	SAO 037822	TYC 3289-2064-1
EU And	SAO 052967	SAO 053004
RS Aqr*	TYC 5201-0192-1	GSC 05201-01479
S Aql	SAO 105808	TYC 1618-1495-1
S Aur	TYC 2411-1188-1	TYC 2411-2135-1
U Boo	TYC 1481-0504-1	GSC 01481-00446
V Boo	SAO 064187	SAO 064166
RW Boo	SAO 064264	SAO 064243
RX Boo	SAO 083321	SAO 083337
U Cam	SAO 012900	SAO 012866
RS Cnc	SAO 061288	SAO 080745
V CVn	SAO 044591	SAO 044551
WZ Cas	SAO 021005	SAO 020984
W Cyg	SAO 050934	SAO 050936
RU Cyg	SAO 033682	TYC 3967-1406-1
RS Cyg	SAO 069635	TYC 3151-1755-1
TT Cyg	SAO 068626	SAO 068629
V778 Cyg	SAO 032689	SAO 032728
U Del	SAO 106396	SAO 106603
RY Dra	SAO 016018	SAO 015879
UX Dra	SAO 009296	SAO 009392
X Her	SAO 045895	SAO 045850
SX Her	SAO 084212	SAO 084221
RT Hya*	SAO 135967	SAO 135972
RS Lac	TYC 3211-0412-1	TYC 3211-0746-1
U LMi	SAO 061743	GSC 02508-00679
X Lib*	TYC 6196-0048-1	TYC 6197-0774-1
X Mon	SAO 133965	SAO 133930
RV Peg*	SAO 072329	GSC 02734-01339
S Per	SAO 023236	TYC 3698-0184-1
U Per	SAO 022853	TYC 3688-0234-1
RW Sgr*	TYC 6304-0290-1	TYC 6300-2225-1
W Tau	TYC 1265-0016-1	TYC 1265-1072-1
Z UMa	SAO 028201	SAO 028193
RZ UMa	SAO 014492	TYC 4132-0395-1
ST UMa	SAO 043786	SAO 043764
R UMi	SAO 008567	SAO 008520
SW Vir	SAO 139201	SAO 139218

* Limited data.

brightness. To produce a cleaner data set for further analysis the reduced data were represented by a manually guided spline fit; this is the smooth curve in the light curve figures.

Low-resolution spectroscopic measurements have also been made of these same stars, including a long (1997–2013) series of observations of RS Cyg and more limited, but continuing (2000–present), spectroscopic monitoring of the other stars. These data were acquired with a conventional Cassegrain spectrograph and either a Reticon photodiode array detector for the RS Cyg series or a CCD camera for the continuing monitoring project. The spectrograph was configured in its lowest-resolution mode, in which resolution is sacrificed for the sake of increased spectral coverage: about 6,000 Å for the Reticon detector and about 3,000 Å for the CCD. In both cases the dispersion is about 6.5 Å/pixel and the resolution is about 10Å. The RS Cyg spectra were corrected for instrumental effects and atmospheric extinction, but no flux correction has been done so far. The reduction of the spectra from the monitoring project is in progress.

4. Results

Representative examples of the photometric data are shown in Figures 1 and 2. The stars in this program exhibit a wide variety of light curve shapes. For most of the stars the variation in B–V is small compared to the variation in V, but all the carbon stars show large variations in B–V that indicate that the blue end of the spectra of these stars varies by about twice as much as does the red end.

A number of the stars did undergo quiescent episodes, revealing small oscillations that were not apparent in the AAVSO data. An example is shown in Figure 2. As an aside, data sets like this are subject to a constraint that is similar to the quantum mechanical Uncertainty Principle (which is itself just a manifestation of wave behavior). For a given set of data the magnitude resolution can often be improved by binning in time, but at the expense of temporal resolution; the product of magnitude uncertainty and time uncertainty can be imagined to be equal to a constant, even if that is not the precise mathematical relationship. For instrumentally acquired data that constant is much smaller than for visual data, which is why we have enough magnitude resolution to be able to see the relatively rapid time variation in a low amplitude signal during the quiescent episodes. Nevertheless, the extensive data accumulated by visual observers were critical in the conception of the project described here and continue to be vital to many other research projects.

In some cases the frequency during the quiescent episode is about the same as that when the amplitude is large, but in most cases it is higher. It was data like these, and specifically the "quiescent" to "normal" frequency ratio, that led to our earlier conclusion that the semiregular variables were likely to be overtone pulsators (Cadmus *et al.* 1991). As we have observed a greater number of quiescent episodes in a variety of stars we have found that these frequency ratios tend to be near integers.

An obvious thing to do with light curves like these is to subject them to Fourier analysis in order to determine what frequencies are present and with what strengths, although

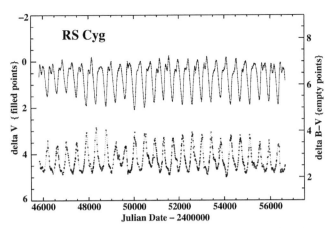

Figure 1. The V light curve (above) and B–V color curve (below) for RS Cyg. This is a carbon star that shows the characteristic large variations in B–V.

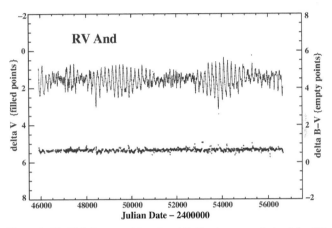

Figure 2. The V light curve (above) and B–V color curve (below) for RV And. This star exhibits quiescent episodes. The small variation in B–V is characteristic of the non-carbon stars.

this exercise is fraught with peril, especially with complex light curves like these. We have nevertheless done that in two different ways. First, we did a straightforward Fourier analysis of each complete light curve. The values of the ratios of the frequencies of the peaks above the dominant peak to that of the dominant peak itself clump near integer values, a harmonic relationship that is characteristic of the Fourier spectra of periodic waveforms and similar to the distribution of the quiescent episode frequency ratios described above. This shows that the frequencies that appear during the quiescent episodes are among those that characterize the light curve in general, but they manage to exist independently, not simply as a component of a shape. We have used the information on the multiple frequencies present in the light curves to construct a plot of the period-luminosity relationships for the various components, resulting in a set of sequences much like that obtained for LMC data (Wood 2000).

Our second approach was a time-dependent Fourier analysis, in which segments of the light curve are analyzed separately and those results are combined to make a frequency vs. time "map." (Similar kinds of maps based on wavelet analysis can be generated using the WWZ function in the AAVSO's vstar package.) This approach clearly shows how a large-amplitude component can temporarily fade, leaving behind a lower-amplitude, higher-frequency component and giving the impression of a mode switch. The temporal variation of the strengths of the frequency components is often quite complex.

Although much of the work in this field has been focused on the relationship between the frequency composition of light curves and the pulsational modes of the stars, there has also been interest in the possible role of chaotic processes in determining the shapes of some light curves. Grinnell photometric data have been compared with theory to show that chaotic processes may be involved (Buchler *et al.* 2004). There is probably more to be done in this area.

Spectroscopic monitoring of all the stars in this program is an ongoing project, but we have made an especially intensive effort to obtain a large set of spectra for RS Cygni; an example is shown in Figure 3. The light curve of this star has distinctive dips in most of its maxima (see Figure 1) and spectra might reveal their cause. The spectra were arranged to make a plot of intensity vs. wavelength and phase but no obvious correlation between line strengths and the photometric dip was immediately apparent. This may indicate that dust is responsible and work is continuing on this puzzle.

Long term, high resolution spectroscopic monitoring of these stars would be valuable, especially to provide better information on atmospheric motions. We are currently developing a fiber-fed radial velocity spectrometer to be used for that purpose.

5. Discussion and future directions

The results described here are preliminary and much remains to be done, both observationally and theoretically. The ability to observe the detailed behavior of stars during quiescent episodes appears to be a useful way to constrain the interpretation of Fourier spectra and other results. In addition, there is certainly science to be extracted from the data for the stars that have not yet experienced quiescent episodes. The RS Cyg spectra have revealed an interesting aspect of the behavior of this star but the final interpretation is not yet in hand.

When this project began thirty years ago we were in a position to make observations that were beyond the reach of most AAVSO observers, but that situation has changed dramatically. Now many of those observers, as well as automated photometric observing programs (Percy *et al.* 2001, for example), are able to do the sort of photometric work that we have been doing and it makes sense for Grinnell to focus more on spectroscopic monitoring that is needed but less easily done. However, continuation of the sort of photometry that we have been doing is very desirable for two reasons. First, the longer a high quality light curve becomes, the more useful it is because it may provide context for newly observed phenomena and provide better tests of chaos models. Second, the spectroscopic work that we are doing will be more useful if there are good light curves to accompany it. We would welcome the participation of AAVSO observers in this effort.

6. Acknowledgements

It is a pleasure to thank Lee Anne Willson and many Grinnell College undergraduate students for their assistance in these projects. In addition, the insight of Janet Mattei was essential. Financial support was provided by the National Science Foundation, Research Corporation, the American Astronomical Society, Sigma Xi, and Grinnell College. The comments provided by a referee were helpful and appreciated.

References

Berlioz-Arthaud, P. 2003, *Astron. Astrophys.*, **397**, 943.
Buchler, J. R., Koláth, Z., and Cadmus, R. R., Jr. 2004, *Astrophys. J.*, **613**, 532.
Cadmus, R. R., Jr., Willson, L. A., Sneden, C., and Mattei, J. A. 1991, *Astron. J.*, **101**, 1043.
Cummings, I. N., Hearnshaw, J. B., Kilmartin, P. M., and Gilmore, A. C. 1999, in *Precise Stellar Radial Velocities*, eds. J. B. Hearnshaw and C. D. Scarfe, ASP Conf. Ser., 185, Astronomical Society of the Pacific, San Francisco, 204.
Gerasimovič, B. P. 1928, *Proc. Natl. Acad. Sci.*, **14**, 963.
Ireland, M. J., Scholz, M., and Wood, P. R. 2011, *Mon. Not. Roy. Astron. Soc.*, **418**, 114.
Kiss, L. L., Szatmáry, K., Cadmus, R. R., Jr., and Mattei, J. A. 1999, *Astron. Astrophys.*, **346**, 542.
Kiss, L. L., Szatmáry, K., Szabó, G., and Mattei, J. A. 2000, *Astron. Astrophys., Suppl. Ser.*, **145**, 283.
Lebzelter, T., Kiss, L. L., and Hinkle, K. H. 2000, *Astron. Astrophys.*, **361**, 167.
Mattei, J. A., Foster, G., Hurwitz, L. A., Malatesta, K. H., Willson, L. A., and Mennessier, M.-O. 1997, in *Proceedings of the ESA Symposium "Hipparcos-Venice '97"*, ESA SP-402, ESA Publ. Div., Noordwijk, The Netherlands, 269.
Nicholls, C. P., Wood, P. R., Cioni, M.-R. L., and Soszyński, I. 2009, *Mon. Not. Roy. Astron. Soc.*, **399**, 2063.
Payne-Gaposchkin, C. 1954, *Ann. Harvard Coll. Obs.*, **113**, 191.
Percy, J. R., and Abachi, R. 2013, *J. Amer. Assoc. Var. Star Obs.*, **41**, 193.
Percy, J. R., Besla, G., Velocci, V., and Henry, G. W. 2003, *Publ. Astron. Soc. Pacific*, **115**, 479.
Percy, J. R., and Polano, S. 1998, in *A half-Century of Stellar Pulsation Interpretations*, eds. P. A. Bradley and J. A. Guzik, ASP Conf. Ser., 135, Astronomical Society of the Pacific, San Francisco, 249.

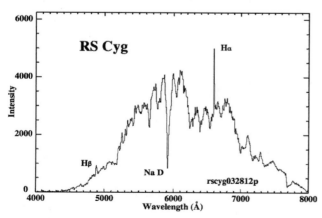

Figure 3. The spectrum of RS Cyg. Note the strong Na absorption line and the Balmer emission lines. The overall shape of the spectrum is instrumental; no flux correction has been applied.

Percy, J. R., and Tan, P. J. 2013, *J. Amer. Assoc. Var. Star Obs.*, **41**, 1.

Percy, J. R., Wilson, J. B., and Henry, G. W. 2001, *Publ. Astron. Soc. Pacific*, **113**, 983.

Perrin, G., Coudé du Foresto, V., Ridgway, S. T., Mennesson, B., Ruilier, C., Mariotti, J.-M., Traub, W. A., and Lacasse, M. G. 1999, *Astron. Astrophys.*, **345**, 221.

Soszyński, I., Dziembowski, W. A., Udalski, A., Kubiak, M., Szymański, M. K., Pietrzyński, G., Wyrzykowski, Ł., Szewczyk, O., and Ulaczyk, K. 2007, *Acta Astron.*, **57**, 201.

Soszyński, I., Wood, P. R., and Udalski, A. 2013a, *Astrophys. J.*, **779**, 167.

Soszyński, I., *et al.* 2013b, *Acta Astron.*, **63**, 21.

Stello, D., *et al.* 2014, *Astrophys. J. Lett.*, **788**, L10.

Stothers, R. B. 2010, *Astrophys. J.*, **725**, 1170.

Willson, L. A. 1976, *Astrophys. J.*, **205**, 172.

Willson, L. A., and Marengo, M. 2012, *J. Amer. Assoc. Var. Star Obs.*, **40**, 516.

Wood, P. R. 1974, *Astrophys. J.*, **190**, 609.

Wood, P. R. 1995, in *Astrophysical Applications of Stellar Pulsation*, eds. R. S. Stobie and P. A. Whitelock, ASP Conf. Ser., 83, Astronomical Society of the Pacific, San Francisco, 127.

Wood, P. R. 2000, *Publ. Astron. Soc. Australia*, **17**, 18.

Wood, P. R., Olivier, E. A., and Kawaler, S. D. 2004, *Astrophys. J.*, **604**, 800.

Wood, P. R., *et al.* 1999, in *Asymptotic Giant Branch Stars*, eds. T. Le Bertre, A. Lèbre, and C. Waelkens, Astronomical Society of the Pacific, San Francisco, 151.

Xiong, D. R., and Deng, L. 2007, *Mon. Not. Roy. Astron. Soc.*, **378**, 1270.

Photometric Analyses and Spectral Identification of the Early-Spectral Type W UMa Contact Binary V444 Andromedae

Ronald G. Samec
Faculty Research Associate, Pisgah Astronomical Research Institute, One Pari Drive, Rosman, NC 28772; and Emmanuel College, 181 Springs St., PO Box 129, Franklin Springs, GA 30639; ronaldsamec@gmail.com

Russell Robb
University of Victoria, Department of Physics and Astronomy, P.O. Box 3055, Station CSC, Victoria, BC V8W 3P6, Canada; and Guest Observer, Dominion Astrophysical Observatory

Danny R. Faulkner
University of South Carolina, 476 Hubbard Drive, Lancaster, SC 29720

Walter Van Hamme
Florida International University, Department of Physics, 11200 SW 8th Street, Miami, FL 33199

Received October 31, 2014; revised December 9, 2014; accepted December 9, 2014

Abstract Presented here are the first precision $UBVR_cI_c$ light curves, synthetic light curve solutions, a period study, and spectrum for V444 Andromedae, an F0V contact W UMa binary. Observations were taken with the Lowell Observatory 0.81-m reflector during 28 through 30 September 2012. From our period study we determined a linear ephemeris showing that the period has been stable over the past 9.6 years (~7,500 orbits). The period during this interval is 0.46877942 d. After a q-search, the lowest residual mass ratio was found to be 0.48 with a Roche-lobe fill-out of nearly 51%. Despite its rather high temperature, 7,300 K, two magnetic spots were modeled on the primary component, a 10° radius equatorial dark spot with a T-factor of 0.88 and a 23° radius near-polar hot spot of T-factor 1.10. The component temperature difference was only ~80 K. These parameters point to a mature, early type, W UMa binary.

1. Introduction

W UMa variable stars are thought to reach contact starting from a detached stage of perhaps a 3- to 5- orbital days period (Yildiz 2014) and slowly reach contact and remain so over eons, steadily losing angular momentum via magnetic braking. Short period, P < 1/2 d, contact binaries are believed to be among the oldest stars in the galaxy (Yildiz 2014). They continue to coalesce with the secondary component losing mass to the primary in a contact Roche Lobe configuration, while steadily becoming more extreme in fill-out and in mass ratio, until the binary becomes a single star in a rapid, observationally-rare event. Ultimately, these binaries become an F to A–type fast rotating, single, blue straggler-like star (Jiang et al. 2014). Our studies have followed the stages through most of the cycle (Yakut and Eggleton 2005). This article gives a first analysis of such a binary, V444 Andromedae, apparently maturing in its contact stage.

2. Observational history

V444 Andromedae (MisV1097, GSC 02808-00139, USNO-A2.0 1275.00747331) is a MISAO Project Variable discovered by Seiichi Yoshida, Nobuo Ohkura, Okayama, Japan, and Ken-ichi Kadota (http://www.aerith.net/misao/data/misv.cgi?1097). Ondrej Pejcha (2005) classified it as a blue W UMa-type (EW/KE-type) eclipsing variable (Kazuhiro Nakajima's observations gave an unfiltered CCD light curve and estimated it to vary from 13.04 to 13.74 magnitudes, with a period of 0.4688 day (Nakajima et al. 2005). Times of minima observed by Ondrej Pejcha have been published (2005). Five times of minimum light are given in *Variable Star Bulletin No. 42* (Nagai 2004). V444 And is listed in the 78th *Name-List of Variable Stars* (Kazarovets et al. 2006) with position R.A. 01^h 15^m 28.7^s Dec. +41° 19' 59", and magnitude range of 13.0–13.7. It was classified as an EW binary. Finally, two times of minimum light were observed by Lewandowski et al. (2007).

3. The present observations

The V444 And system was observed as a part of our student/professional collaborative studies of interacting binaries taken from data at the NURO (National Undergraduate Research Observatory). The observations were taken by Dr. Ron Samec, Dr. Danny Falkner, Sharyl Monroe, and Heather A. Chamberlain. The reductions and analyses were done by Travis Shebs and Dr. Samec.

Our UBVRI light curves of V444 And were taken with the Lowell 31-inch reflector in Flagstaff, Arizona, with a CRYOTIGER cooled (< –100° C) NASACAM and a 2K × 2K chip (Plate Scale = 0.515 arcsec/pixel, http://www.nuro.nau.edu/specs.htm) and standard Johnson-Cousin BVR_cI_c filters. They were observed on 28, 29, 30 September 2012. The light curves were subjected to synthetic light curve modeling. The observations included 48 in U filter, 128 in B, R_c, and I_c-filters,

Table 1. The variable (V), comparison (C), and check (K) stars in this study.

Star	Name	R.A. (2000) h m s	Dec. (2000) ° ′ ″	V	Source*
V	V444 And GSC 2808 0139	01 15 28.72	+41 19 58.9	12.86	IRCS
C	GSC 2808 0101	01 15 52.759	+41 22 15.56	12.93	Guide 9
K	GSC 2808 0111	01 15 24.924	+41 18 43.23	12.13	Guide 9

*Guide 9 (Project Pluto 2012).

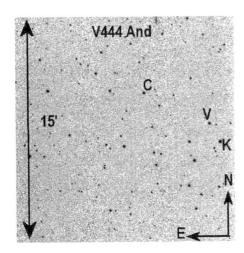

Figure 1. Finding chart for V444 And (V), comparison (C) and check (K).

Table 2. O–C Residuals calculated from Equation 1.

	Epoch JD 2400000+	Cycle	O–C	Reference
1	52685.2892	–7497.5	0.0011	Nakajima et al. 2005
2	52911.0059	–7016.0	0.0005	Nakajima et al. 2005
3	52912.1761	–7013.5	–0.0013	Nakajima et al. 2005
4	52914.0532	–7009.5	0.0007	Nakajima et al. 2005
5	52914.2861	–7009.0	–0.0008	Nakajima et al. 2005
6	53966.6965	–4764.0	–0.0002	Diethelm 2007
7	53966.9314	–4763.5	0.0003	Diethelm 2007
8	54097.2513	–4485.5	–0.0005	Diethelm 2007
9	56198.7907	–2.5	0.0008	Present observations
10	56199.0239	–2.0	–0.0004	Present observations
11	56199.9616	0.0	–0.0002	Present observations

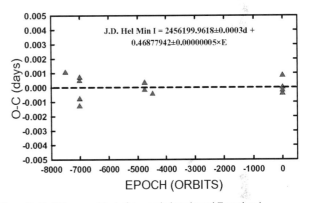

Figure 2. O–C linear residuals from period study and Equation 1.

and 127 in V, individual CCD observations. The standard errors of a single observation were 1.3 mmag in U, 7 mmag B and I, 4 mmag V, 4 mmag R_c. Typical exposure times were 300s in U, 75s in B, 50s in V, and 35s in R_c and I_c. Images were calibrated in a standard way using biases, 300-sec darks and UBVRI sky flats with AIP4WIN (Berry and Burnell 2011, CFITSIO package developed by NASA). The stars in this study (Table 1) are shown on the finding charts in Figure 1. The UBVRI observations in delta magnitudes, in the sense of V–C, are given online at http://smileycontrol.pari.edu/smileydata/Users/Samec/V444And .

4. Period determination

Three new times of minimum light were determined from our observations. These include:

$$JD\ Hel\ Min\ I = 2456199.0239 \pm 0.0011,$$
$$2456199.9616 \pm 0.0015,$$

and

$$JD\ Hel\ Min\ II = 2456198.7907 \pm 0.0005.$$

From all available timings, an improved linear ephemeris was determined:

JD Hel Min I =
$2456199.9618 \pm 0.0003d + 0.46877942 \pm 0.00000005 \times E$ (1)

The O–C residuals calculated using Equation 1 are given along with all available timings and their O–C residuals in Table 2. The plot of residuals is given in Figure 2.

5. Spectral type, temperature

Robb observed the spectrum of V444 And at Dominion Astrophysical Observatory (DAO) with the 1.8-m telescope at a dispersion of 60 Å/mm. The mid-time of exposure was 23 Jul 2013 10:20:03 UT and lasted 999 seconds which was at MJD 2456496.4181, corresponding to phase 0.48. The strengths of all lines indicate a F0V ± 1 (7,300 K) spectral type. The spectrum and the comparison spectrum of ρ Gem are given in Figure 3. The B–V for an F0V Star is ~0.3. From the AAVSO Photometric All-Sky Survey (APASS; Henden et al. 2013), B–V = 0.42. This gives a reddening of E(B–V) = 0.12.

6. Light curve characteristics

Phased light curves using Equation (1) to fold the light curves are given in Figures 4a, 4b, and 4c.

The ΔU to ΔI$_c$ magnitude curve amplitudes are from 0.72 to 0.63 magnitude, respectively. Whereas, the O'Connell (MAXII-MAXI) effect is small as compared to the errors, it does

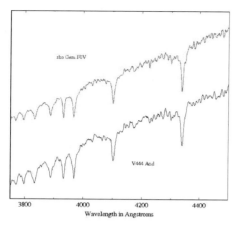

Figure 3. DAO spectra of V444 And and comparison star.

Table 3. Light curve characteristics.

Filter		Magnitude	
		Max. I	Max. II
	Phase	0.25	0.75
ΔU		−0.305±0.005	−0.285±0.013
ΔB		−0.278±0.007	−0.258±0.005
ΔV		−0.201±0.011	−0.179±0.003
ΔR		−0.149±0.013	−0.133±0.002
ΔI		−0.069±0.012	−0.064±0.004
Δ(U–B)		−0.027±0.012	−0.027±0.018
Δ(B–V)		−0.077±0.018	−0.078±0.007
Δ(R–I)		−0.072±0.025	−0.054±0.006
		Min. II	Min. I
	Phase	0.5	0.0
ΔU		0.360 —	0.410 —
ΔB		0.371±0.006	0.400±0.028
ΔV		0.432±0.003	0.475±0.005
ΔR		0.485±0.004	0.515±0.011
ΔI		0.560±0.008	0.561±0.011
Δ(U–B)		−0.011±0.006	0.010±0.028
Δ(B–V)		−0.061±0.009	−0.075±0.033
Δ(R–I)		0.546±0.012	0.589±0.022

	Min. I–Max. I	Min. II–Max. I	Min. I–Min. II
ΔU	0.715±0.005	0.020±0.018	0.050±0.000
ΔB	0.678±0.035	0.021±0.012	0.029±0.034
ΔV	0.676±0.016	0.022±0.014	0.044±0.008
ΔR	0.664±0.024	0.017±0.015	0.030±0.015
ΔI	0.630±0.023	0.006±0.015	0.001±0.019
Δ(U–B)	0.037±0.040	−0.001±0.029	0.021±0.034
Δ(B–V)	0.002±0.051	−0.001±0.025	−0.014±0.042
Δ(R–I)	0.661±0.047	0.017±0.030	0.044±0.034

signal the presence of spots, averaging about 0.02 magnitude. The depths of eclipse are very nearly equal to within 0–5%, so the temperatures of the binary components are nearly identical. This is not unexpected for a high fill-out mature W UMa binary. The magnitude differences of the light curve characteristics at quadratures are given in Table 3.

7. Synthetic light curve modeling

Each filter's light curve was fitted, individually, with BINARY MAKER 3.0 (Bradstreet and Steelman 2002) using standard convective parameters and limb darkening coefficients from tabled values dictated by the spectral type. In these models we used both under-luminous and over-luminous spots to fit the asymmetries in the curves. Using our averaged values from BINARY MAKER and the primary component temperature from our spectrum (7,300 K), we proceeded to compute a UBVRI simultaneous five color light curve solution with the Wilson Devinney Program (Wilson and Devinney 1971; Wilson 1990, 1994; Van Hamme and Wilson 1998) includes Kurucz stellar atmospheres rather than black body. We used two-dimensional limb-darkening coefficients and a detailed reflection treatment in our modeling. Our fixed inputs included standard convective parameters: gravity darkening, g, = 0.32 and albedo value of 0.5. We used Mode 3 in our analysis (the contact configuration). Since the eclipses were not total, a q-search was prescribed. There is no third light in our solution. Only small negative and

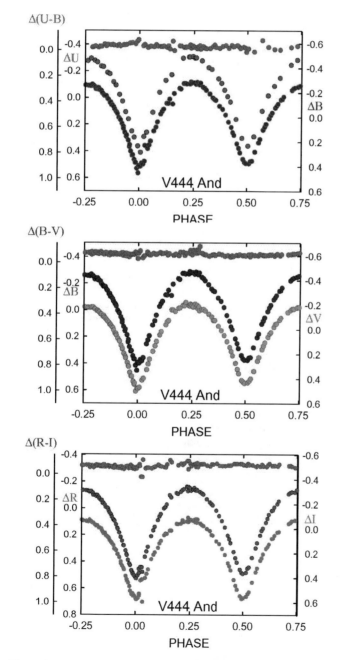

Figures 4a (top), 4b (middle), 4c (bottom). Phased ΔU, ΔB, ΔV, ΔR$_c$, and ΔI$_c$ differential magnitude light curves and Δ(U–B), Δ(B–V), Δ(R$_c$–I$_c$) color curves.

negligible values resulted when it was included in the adjustable parameters, indicating that no third body of appreciable brightness contributes to the overall light of the system. So after a sufficient q-search, the lowest residual mass ratio was found to be 0.48. The residual vs. mass ratio plot is given in Figure 5. Adjustable parameters include those accompanied by errors (see Table 4), the inclination, i, the temperature of the secondary component, T_2, the potential, Ω, the mass ratio, $q = M_2/M_1$, the normalized flux (at 4π) in each wavelength, L, the phasing ephemeris, JD_o, the period P, and the four spot parameters. Our complete solution is given in Table 4 and the curves are shown in Figures 6a and 6b, where the solution is overlying the normalized flux light curves. The Roche lobe surfaces arising from the calculation are displayed in Figures 7a, 7b, 7c, and 7d. We also calculated a nonspotted solution for comparison. Note that the sum of square residuals was 0.425 for the spotted solution and 0.601 for the unspotted one (41% larger). Finally, we calculated a radiative solution, with radiative parameters, gravity darkening, g, = 1.00 and albedo value of 1.0. The solution is given in Table 4. The main differences are the light-related phenomena and the fill-out which is substantial but smaller. The sum of square residuals was 0.425 for the convective solution and 0.579 for the radiative one (36% larger). So, the convective solution was best.

8. Discussion

V444 And is an early-type contact W UMa system. It is of F0V spectral type. The period has been constant over the past 9.6 years. The orbital period during this interval is 0.46877942 d. The lowest residual mass ratio is found to be 0.48 and the fill-out is rather large, some 51%. Despite its temperature (7,300 K), two magnetic spots were modeled on the primary component, a small 10° radius equatorial dark spot, and a moderate-sized 23° radius hot polar spot. The inclination was above 80° so its eclipses are nearly total, with only 1–2% of the light contributed by the secondary component at phase 0.5. The component temperature difference is only ~80 K, so despite the disparate mass ratio, the components are in good thermal contact. Since the binary is magnetically active, we expect this system will eventually become a fast spinning early A-type star. The process of coalescence is caused by magnetic braking due to stellar winds leaving the system via stiff dipole magnetic field lines. We note that radial velocity curves are needed to confirm the mass ratio and to determine absolute mass values. In addition, this system should be patrolled for the next ten years or so to determine possible changes in the period behavior of the system.

9. Acknowledgements

We wish to thank NURO for their allocation of observing time and Bob Jones University for their support of our observing runs over the past eighteen years. We also wish to think Pisgah Astronomical Research Institute for appointing the primary author as a Faculty Research Associate. This research was made possible through the use of the AAVSO Photometric All-Sky Survey (APASS), funded by the Robert Martin Ayers Sciences Fund.

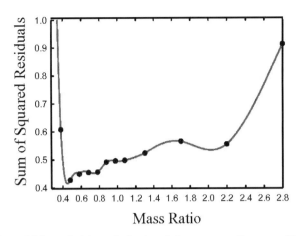

Figure 5. Mass ratio (q) search, fixed q-solutions vs. sum of square residuals to determine the best q-value.

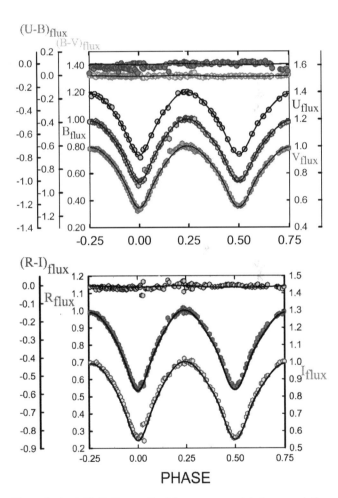

Figure 6a (top). U, B, V normalized flux curves overlain by our solution.
Figure 6b (bottom). R_c, I_c normalized flux curves overlain by our solution.

Table 4. V444 And, synthetic light curve solution.

Parameters	Convective Solution	Radiative Solution
$\lambda_U, \lambda_B, \lambda_V, \lambda_R, \lambda_I$ (nm)	360, 440, 550, 640, 790	360, 440, 550, 640, 790
$x_{bol1,2}, y_{bol1,2}$	0.642, 0.642, 0.260, 0.260	0.642, 0.642, 0.260, 0.260
$x_{1I,2I}, y_{1I,2I}$	0.517, 0.517, 0.283, 0.283	0.517, 0.517, 0.283, 0.283
$x_{1R,2R}, y_{1R,2R}$	0.604, 0.604, 0.300, 0.300	0.604, 0.604, 0.300, 0.300
$x_{1V,2V}, y_{1V,2V}$	0.683, 0.683, 0.298, 0.298	0.683, 0.683, 0.298, 0.298
$x_{1B,2B}, y_{1B,2B}$	0.782, 0.782, 0.300, 0.300	0.782, 0.782, 0.300, 0.300
$x_{1U,2U}, y_{1U,2U}$	0.778, 0.778, 0.338, 0.338	0.778, 0.778, 0.338, 0.338
g_1, g_2	0.32	1.00
A_1, A_2	0.5	1.0
Inclination (°)	80.85±0.07	79.46±0.06
T_1, T_2 (K)	7300, 7225±2	7300, 7138±3
$\Omega_1 = \Omega_2$	2.692±0.002	2.737±0.002
$q(m_2/m_1)$	0.4803±0.0007	0.478±0.001
Fill-outs: $F_1 = F_2$	50.5±0.3%	33±1%
$L_1/(L_1+L_2)_I$	0.6569±0.0019	0.6699±0.0011
$L_1/(L_1+L_2)_R$	0.6587±0.0014	0.6734±0.0008
$L_1/(L_1+L_2)_V$	0.6604±0.0011	0.6773±0.0006
$L_1/(L_1+L_2)_B$	0.6628±0.0017	0.682±0.001
$L_1/(L_1+L_2)_U$	0.6620±0.0019	0.680±0.001
JD_o (days)	56199.9620±0.0002	56199.9620±0.0002
Period (days)	0.46862±0.00008	0.46862±0.00008
r_1, r_2 (pole)	0.444±0.003, 0.324±0.005	0.435±0.003, 0.314±0.005
r_1, r_2 (side)	0.479±0.004, 0.343±0.006	0.467±0.005, 0.330±0.006
r_1, r_2 (back)	0.520±0.007, 0.401±0.014	0.520±0.007, 0.377±0.012
Spot 1 On STAR 1	Cool Spot	Cool Spot
Colatitude (°)	81.9±1.4	82±4
Longitude (°)	97.4±0.9	97±6
Spot radius (°)	9.9±0.5	10±1
Spot T-factor	0.883±0.004	0.883±0.004
Spot 2 On STAR 1	Hot Spot	Hot Spot
Colatitude (°)	21.1±0.7	17±1
Longitude (°)	358.3±0.9	355±2
Spot radius (°)	23.0±0.3	23.0±0.4
Spot T-factor	1.10±0.01	1.074±0.004
Σres^2	0.42515	0.57855

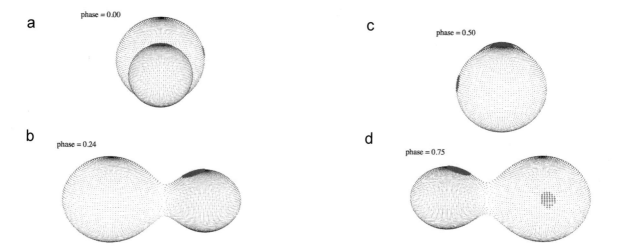

Figures 7a, 7b, 7c, 7d. Roche Lobe surfaces of V444 And.

References

Berry, R., and Burnell, J. 2011, "Astronomical Image Processing for Windows," version 2.4.0, provided with *The Handbook of Astronomical Image Processing*, Willmann-Bell, Richmond, VA.

Bradstreet, D. H., and Steelman, D. P. 2002, *Bull. Amer. Astron. Soc.*, **34**, 1224.

Diethelm, R. 2007, *Inf. Bull. Var. Stars*, No. 5781, 1.

Henden, A. A., *et al.* 2013, AAVSO Photometric All-Sky Survey, data release 7 (http://www.aavso.org/apass).

Jiang, D., Han, Z., and Li, L. 2014, *Mon. Not. Roy. Astron. Soc.*, **438**, 859.

Kazarovets, E. V., Samus, N. N., Durlevich, O. V., Kireeva, N. N., and Pastukhova, E. N. 2006, *Inf. Bull. Var. Stars*, No. 5721, 1.

Lewandowski, M., Niedzielski, A., and Maciejewski, G. 2007, *Inf. Bull. Var. Stars*, No. 5784, 1.

Nagai, K. 2004, *Var. Star Bull.* (Japan), No. 42, 1.

Nakajima, K., Yoshida, S., Ohkura, N., and Kadota, K. 2005, *Inf. Bull. Var. Stars*, No. 5600, Report No. 7.

Pejcha, O. 2005, *Inf. Bull. Var. Stars*, No. 5645, 1.

Van Hamme, W., and Wilson, R. E. 1998, *Bull. Amer. Asron. Soc.*, **30**, 1402.

Wilson, R. E. 1990, *Astrophys. J.*, **356**, 613.

Wilson, R. E. 1994, *Publ. Astron. Soc. Pacific*, **106**, 921.

Wilson, R. E., and Devinney, E. J. 1971, *Astrophys. J.*, **166**, 605.

Yakut, K., and Eggleton, P. P. 2005, *Astrophys. J.*, **629**, 1055.

Yildiz, M. 2014, *Mon. Not. Roy. Astron. Soc.*, **437**, 185.

The Curious Case of ASAS J174600-2321.3: an Eclipsing Symbiotic Nova in Outburst?

Stefan Hümmerich
Stiftstr. 4, Braubach, D-56338, Germany; American Association of Variable Star Observers (AAVSO), Cambridge, MA; Bundesdeutsche Arbeitsgemeinschaft für Veränderliche Sterne e.V. (BAV), Berlin, Germany; stefan.huemmerich@gmail.com

Sebastián Otero
Olazabal 3650-8 C, Buenos Aires, 1430, Argentina; American Association of Variable Star Observers (AAVSO), Cambridge, MA

Patrick Tisserand
Sorbonne Universités, UPMC Univ Paris 06, UMR 7095, Institut d'Astrophysique de Paris, F-75005 Paris, France; CNRS, UMR 7095, Institut d'Astrophysique de Paris, 98 bis Boulevard Arago, F-75014 Paris, France

Klaus Bernhard
Kafkaweg 5, Linz, 4030, Austria; American Association of Variable Star Observers (AAVSO), Cambridge, MA; Bundesdeutsche Arbeitsgemeinschaft für Veränderliche Sterne e.V. (BAV), Berlin, Germany

Received October 15, 2014; revised November 25, 2014; accepted December 15, 2014

Abstract The star ASAS J174600-2321.3 was found to exhibit peculiar photometric variability (conspicuous brightening of ~4 magnitudes (V), fast luminosity declines, intrinsic pulsations). It was rejected as an RCB candidate in recent investigations on spectroscopic grounds. We have collected and present all available data from public sky surveys, photometric catalogues, and the literature. From an analysis of these data, we have identified ASAS J174600-2321.3 as a long-period eclipsing binary (P_{orb} = 1,011.5 days). The primary star, which is probably a white dwarf, is currently in outburst and exhibits the spectral characteristics of a reddened, early F-type supergiant; the secondary star is a giant of spectral type late M. We discuss the possible origin of the observed brightening, which is related to the primary component. ASAS J174600-2321.3 is most certainly an eclipsing symbiotic binary—probably a symbiotic nova of GCVS type NC—that is currently in outburst. However, further photometric and spectroscopic data are needed to confirm this.

1. Introduction

ASAS J174600-2321.3 = EROS2-cg1131n13463 = 2MASS J17460018-2321163, which is situated in Sagittarius at position (J2000) R.A. $17^h 46^m 00.180^s$ Dec. $-23° 21' 16.37''$ (UCAC4; Zacharias et al. 2012), was first published as a variable star in the ASAS Catalog of Variable Stars (ACVS; Pojmański et al. 2005), where it was catalogued as a miscellaneous variable (type "MISC"). It was found to exhibit peculiar photometric variability (conspicuous rise in brightness, pulsations). Consequently, it was investigated as a candidate R Coronae Borealis (RCB) star during several recent investigations (Tisserand et al. 2008; Tisserand et al. 2013) but ultimately rejected on spectroscopic grounds, mainly because of a high abundance of hydrogen. Identifications and coordinates of ASAS J174600-2321.3 are given in Table 1.

We have compiled all available data on this object from various sky surveys, photometric catalogues, and the literature; they are presented in Section 2 and analyzed and interpreted in Section 3. Possible reasons for the rise in magnitude and probable classifications are discussed in Section 4. We conclude in Section 5.

2. Observations

2.1. Multiband photometric data

Before presenting the photometric data, it is important to point out that there is considerable line-of-sight extinction to ASAS J174600-2321.3, which affects the recorded photometric properties of the star. All dust-reddening estimates in this and the following sections are based on the work of Schlafly and Finkbeiner (2011), who employ the colors of stars with spectra in the Sloan Digital Sky Survey (Adelman-McCarthy et al. 2011) and measure reddening as the difference between the

Table 1. Identifications and coordinates of ASAS J174600-2321.3. Positional data were taken from UCAC4.

Identifiers	R.A. (J2000) h m s	Dec. (J2000) ° ' "	Galactic Longitude (°)	Galactic Latitude (°)
ASAS J174600-2321.3 EROS2-cg1131n13463 2MASS J17460018-2321163 USNO-B1.0 0666-0562145	17 46 00.180	–23 21 16.37	2.829	4.814

measured and predicted colors of a star. Their results prefer an $R_V = 3.1$ Fitzpatrick (1999) reddening law. Based on the calculations of Schlafly and Finkbeiner (2011), we estimate an interstellar extinction of $A_V \approx 2.4$ mag. and $E(B–V) \approx 0.78$ mag. for the sky area of our interest. The extinction values in different bandpasses, which have been used for the reddening correction in the present paper, were accessed through the NASA/IPAC Infrared Science Archive (http://irsa.ipac.caltech.edu/applications/DUST/). If not indicated otherwise, the light curves illustrated in this and the following sections are based on non-corrected data.

In order to achieve a long time baseline, we have combined photometric observations from the EROS-2 project (Renault et al. 1998), ASAS-3 (Pojmański 2002), and APASS (Henden et al. 2012). EROS-2 observations were performed with two wide field cameras behind a dichroic cube splitting the light beam into two broad passbands. The so-called "blue" channel (BE, 420-720 nm) overlapped the V and R standard bands; the "red" channel (RE, 620-920 nm) roughly matched the mean wavelength of the Cousins I band. Data from the EROS-2 project have been transformed to Johnson V and Cousins I using Equation (4) of Tisserand et al. (2007):

$$R_{eros} = I_C \qquad B_{eros} = V_J - 0.4(V_J - I_C) \qquad (1)$$

The obtained results are accurate to a precision of 0.1 mag. (Tisserand et al. 2007). Unfortunately, problems with the CCD detectors of the "red" camera resulted in missing R_E data after HJD 2452200. Thus, no contemporaneous R_E measurements were available to transform the B_E values after HJD 2452200 using Equation (1). The missing data have been linearly interpolated in such a way that the resulting data forms a continuity with the preceding EROS-2 and succeeding ASAS-3 measurements. We have achieved this by using Equation (2), which has been derived from the conversion of data before HJD 2452200. As the system brightened considerably during the time of the missing R_E measurements, and the brightening is obviously related to the hot primary component, we have also taken into account the resulting color changes and have adapted the equation accordingly. (As primary star, we define the star which is the seat of the observed brightening and which we propose as possible accretor in a symbiotic binary scenario.)

$$\Delta I_C = 0.57 \times \Delta B_E \qquad (2)$$

As can be seen from the resulting light curve, there is very good agreement between the converted EROS-2 and ASAS-3 data. Furthermore, the derived color index from the resulting data, $(V–I_C)_{eros} = 1.9$ mag. at $V \approx 13.0$ mag., is consistent with the index derived from APASS photometry at maximum brightness, $(V–I_C)_{apass} = 1.5$ mag. at $V \approx 12.2$ mag. We thus feel confident about the applicability of our solution.

The combined light curve of ASAS J174600-2321.3, based mainly on sky survey data from EROS-2, ASAS-3, and APASS, is shown in Figure 1; APASS measurements are presented in Table 2.

During the beginning of EROS-2 coverage, the star's brightness varied slowly around 16.5 mag. (V). A rise in magnitude set in at around HJD 2451240, with the system reaching 15 mag. (V) about 350 days later. After a short plateau and the eclipse at around HJD 2452100, ASAS J174600-2321.3 continued to brighten in V during the rest of EROS-2 and the beginning of ASAS-3 coverage, rising by about three magnitudes in ~500 days. The star reached ~12.2 mag. (V) at about HJD 2452600, around which it has apparently remained up to the present time. Interestingly, the conspicuous brightening is not seen at longer wavelengths; the mean brightness of the corresponding EROS-2 I_C measurements increased by only 0.2 mag., although—as mentioned above—there exists no I_C photometry for the most dramatic part of the brightening between about HJD 2452200 and HJD 2452680 because of the missing R_E measurements after HJD 2452200. In order to derive an approximate amplitude in I_C, we have transformed APASS photometry at maximum using Equation (3) of Jester et al. (2005; especially their Table 1):

$$R - I = 1.00(r - i) + 0.21 \qquad (3)$$

From this, we derive $I_C = 10.6$ mag. at maximum, which indicates a total amplitude of ~2 mag. in this band—less than half of the amplitude in V.

2.2. The observed outburst and a historical light curve

As becomes obvious from Figure 1, the system shows complex photometric variability on different timescales. The most striking feature of the light curve is the already mentioned drastic increase in visual brightness related to the primary component. This outburst has been accompanied by a blueward color evolution. Before the rise in brightness (HJD < 2451112), we derive a mean extinction corrected color index of $(V–I_C)_0 \approx 3.0$ mag., which is consistent with a spectral type of ~mid-M (Ducati et al. 2001). At the last simultaneous V and I_C measurements at HJD 2452200.542—after the system had already brightened by ~1.5 mag. (V)—the color index became bluer, to $(V–I_C)_0 \approx 1.8$ mag. Finally, at maximum visual light,

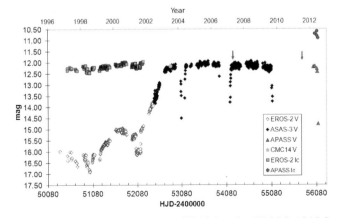

Figure 1. Light curve of ASAS J174600-2321.3, based on EROS-2, ASAS-3, and APASS data. EROS-2 B_E and R_E magnitudes have been transformed to Johnson V and Cousins I, respectively; Cousins I values have been derived from APASS photometry using the equations in Jester et al. (2005); see text for details. Also included is a transformed observation from CMC14 (Evans et al. 2002). Some obvious outliers with a photometric quality flag of D have been removed from the ASAS-3 dataset. The epochs of the spectra are indicated by arrows; magnitudes were not corrected for extinction (see text for details).

Table 2. Photometric measurements of ASAS J174600-2321.3 from APASS. I_c values have been derived from APASS photometry using the equations in Jester et al. (2005).

Passband	HJD–2400000	Magnitude	Error
APASS B	56008.7946	13.208	0.006
	56040.7133	13.189	0.006
	56085.5371	13.347	0.010
	56086.5358	13.476	0.007
	56105.6268	15.992	0.060
APASS V	56008.7942	12.179	0.005
	56040.7128	12.176	0.005
	56085.5366	12.275	0.007
	56086.5353	12.398	0.006
	56105.6263	14.753	0.042
APASS i'	56008.7968	11.098	0.006
	56040.7154	11.274	0.008
	56085.5544	11.342	0.018
	56086.5379	11.368	0.015
	56101.8428	12.445	0.026
	56102.8404	12.562	0.026
	56103.8376	12.609	0.028
	56105.6939	12.502	0.029
APASS g'	56040.7159	12.681	0.004
	56085.5422	12.817	0.006
	56086.5410	12.878	0.004
	56105.6293	15.373	0.035
	56134.6033	16.700	0.109
APASS r'	56040.7141	11.778	0.005
	56085.5379	11.861	0.006
	56086.5366	12.020	0.006
	56105.6276	14.123	0.037
	56134.6016	14.689	0.060
APASS I_c	56040.7128	10.7	0.15
	56085.5366	10.8	0.15
	56086.5353	10.6	0.15
	56105.6263	10.9	0.15

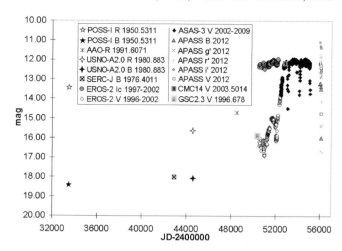

Figure 2. Light curve of ASAS J174600-2321.3, based on data from various historical and contemporary sources, as indicated in the legend. Magnitudes were not corrected for extinction.

Table 3. Measurements of ASAS J174600-2321.3 in various astrometric and photometric catalogues. Magnitudes were not corrected for extinction.

Value	Source	Epoch	Remarks
18.38 (B1)	POSS-I	1950.5311	(B–R) = 4.95
13.43 (R1)	POSS-I	1950.5311	(B–R) = 4.95
18.01 (B2)	SERC-J	1976.4011	
14.69 (R2)	AAO-R	1991.6071	
18.1 (B)	USNO-A2.0	1980.883	(B–R) = 2.8
15.3 (R)	USNO-A2.0	1980.883	(B–R) = 2.8
15.86 (V)	GSC2.3	1996.678	
15.1–15.5 (V)	UCAC3	—	Transformed using 2MASS J–K and APASS r' and V.
12.15 (V)	CMC14	2003.5014	Transformed using APASS r' and V.

the derived color index, $(B-V)_0 \approx 0.2$ mag., is in agreement with a late A- / early F-type supergiant (see sections 3.2 and 3.3).

In order to investigate the long-term photometric variability of ASAS J174600-2321.3, we have constructed a light curve including historical data from various catalogue sources, which is shown in Figure 2. Archival plate measurements and other observations over the years indicate that the star's magnitude has been found between magnitudes 17 and 18 on several occasions on blue-sensitive photographic plates (Table 3), which is about 4 to 5 magnitudes fainter than the observed maximum B magnitude of 13.2 from APASS—a value well beyond any possible error in plate reductions. No bright states seem to have been recorded in the past. However, as there are only a few scattered measurements in different passbands before the onset of the EROS project, the evidence is tentative, at best.

2.3. Spectroscopic observations

During a search for new RCB stars in the EROS-2 database, Tisserand et al. (2008) identified ASAS J174600-2321.3 as a likely candidate on grounds of its RCB-like photometric variability (fast brightness declines; slow, RCB-like recovery at the beginning of ASAS-3 coverage, intrinsic pulsations). A spectrum of the object was obtained using the Dual-Beam Spectrograph (DBS; Rodgers et al. 1988) attached to the 2.3-meter telescope at the Siding Spring Observatory of the Australian National University (ANU). The DBS is a general purpose optical spectrograph, which is permanently mounted at the Nasmyth A focus. The visible waveband is split by a dichroic at around 600 nm and feeds two essentially similar spectrographs, with red and blue optimized detectors. The full slit length is 6.7 arcmin.

The spectrum was taken at maximum visual brightness on HJD ~2454213.9218 (hereafter referred to as the "2007 spectrum") and comprises a wavelength range of $5550 \text{Å} < \lambda < 7500 \text{Å}$ and a 2-pixel resolution of 2 Å. The spectrum is shown in Figure 9.

Using enhanced selection criteria, the star was recovered as an RCB candidate during a more recent search by Tisserand et al. (2013). This renewed the interest in the object and, accordingly, it was re-investigated and a spectrum with broad wavelength coverage was obtained. Spectroscopic observations were performed with the Wide Field Spectrograph (WiFeS) instrument (Dopita et al. 2007) attached to the ANU 2.3-m telescope at the Siding Spring Observatory. WiFeS is an integral field spectrograph permanently mounted at the Nasmyth A focus. It provides a 25" × 38" field with 0.5-arcsec sampling along each of the twenty-five 38" × 1" slitlets. The visible wavelength interval is divided by a dichroic at around 600 nm feeding two essentially similar spectrographs.

The spectrum, which comprises a 2-pixel resolution of 2 Å, was taken at maximum visual brightness on HJD ~2455762.9661 (the "2012 spectrum"); a segment is shown in Figure 10 (5700 Å < λ < 9700 Å), the entire spectrum is given in Figure 11. The epochs of both spectra are indicated in Figure 1.

3. Data analysis and interpretation

We interpret the light curve characteristics (brightening, eclipses, pulsations), the spectral energy distribution, and the spectra as being due to the fact that ASAS J174600-2321.3 is made up of two components. In the following sections, the observed phenomena are discussed in that context.

3.1. Orbital period and eclipse characteristics

Three conspicuous fading episodes were covered to varying degrees by ASAS-3; a fourth one is evident from recent APASS data (Figure 1). From their rapid declines, symmetric recoveries, and strict periodicity, we conclude that these events represent deep eclipses with a period of P_{orb} = 1,011.5 days and an eclipse duration of 111 days, which corresponds to ~11% of the period. No trace of a secondary eclipse could be found. A phase plot illustrating the deep eclipses is shown in Figure 3. The following elements have been derived:

$$\text{HJD (Min I)} = 2456142 + 1011.5 \times E \quad (4)$$

We have searched for additional eclipses of low amplitude in EROS-2 data (i.e., before the most dramatic part of the brightening). There is another eclipse at around HJD 2452100 ("the 2001 eclipse") which exhibits evidence of a pronounced time of totality, during which the primary star is completely hidden by the eclipsing body. It is interesting to note that the pulsational variability—which will be discussed in section 3.4—continues during the observed time of totality (Figure 4).

No other clearly defined eclipses could be found. However, there is a curious, "V-shaped" fading near the expected time of eclipse at around HJD 2451050 ("the 1998 fading"; Figure 5). This is likely not an actual eclipse (it occurs 76.5±25 days before the computed time of minimum as determined by

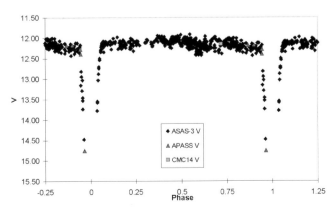

Figure 3. Phase plot of ASAS J174600-2321.3, based on ASAS-3, APASS V, and transformed CMC14 data, and folded with the ephemeris given in Equation (4). Only data > HJD 2452700 have been considered for the phase plot.

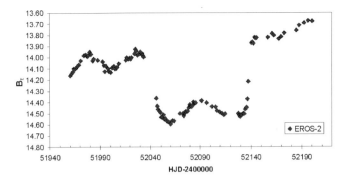

Figure 4. Close-up view of the 2001 eclipse. In order to preserve a maximum number of datapoints, the plot has been based on EROS-2 instrumental B_E magnitudes. Note the continuance of pulsations during the time of totality.

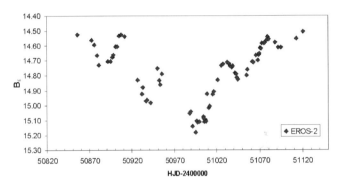

Figure 5. Close-up view of the 1998 fading. In order to preserve a maximum number of datapoints, the plot has been based on EROS-2 instrumental B_E magnitudes.

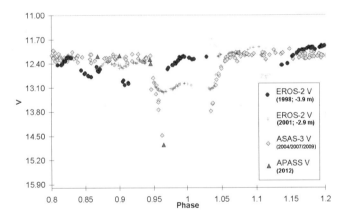

Figure 6. Phase plot of ASAS J174600-2321.3, illustrating all primary eclipses; also included is the 1998 fading episode. The plot has been based on EROS-2, ASAS-3, and APASS V data and folded with the elements given in Equation (4). EROS-2 B_E magnitudes have been transformed to Johnson V (see text for details). In order to facilitate comparison, the resulting EROS-2 V magnitudes have been shifted by the indicated amounts to match V magnitudes from ASAS-3 and APASS.

Table 4. Recent photometric measurements of ASAS J174600-2321.3. Catalogue data have been accessed through VizieR and corrected for interstellar extinction using the estimates of Schlafly and Finkbeiner (2011), which have been accessed through the NASA/IPAC Infrared Science Archive (http://irsa.ipac.caltech.edu/applications/DUST/). The listed error margins are taken directly from the corresponding catalogues.

Passband	Source	Magnitude (Error)
B	APASS (max. vis. light)	9.99 (±0.02)
B	*APASS ($\varphi_{orb} \sim 0.964$)*	*12.78 (±0.02)*
V	APASS (max. vis. light)	9.75 (±0.00)
V	*APASS ($\varphi_{orb} \sim 0.964$)*	*12.33 (±0.00)*
g'	APASS (max. vis. light)	9.69 (±0.01)
g'	*APASS ($\varphi_{orb} \sim 0.964$)*	*12.39 (±0.01)*
r'	APASS (max. vis. light)	9.71 (±0.01)
r'	*APASS ($\varphi_{orb} \sim 0.964$)*	*12.06 (±0.01)*
i'	APASS (max. vis. light)	9.65 (±0.13)
i'	*APASS ($\varphi_{orb} \sim 0.964$)*	*11.07 (±0.13)*
J	2MASS	9.27 (± 0.022)
H	2MASS	8.35 (± 0.044)
K_s	2MASS	8.07 (± 0.024)
[3.6]	GLIMPSE	7.52 (± 0.033)
[4.5]	GLIMPSE	7.60 (± 0.040)
[5.8]	GLIMPSE	7.48 (± 0.039)
[8.0]	GLIMPSE	7.37 (± 0.019)
[3.4]	WISE	7.57 (± 0.026)
[4.6]	WISE	7.62 (± 0.020)
[12]	WISE	7.60 (± 0.022)
[22]	WISE	7.33 (± 0.235)

Color Index	Remark
$(V-I_C)_0 \approx 3.0$	mean value at minimum visual light (HJD < 2451112.53)
$(B-V)_0 \approx 0.2$	epoch: HJD 2456008.7942 (maximum visual light)
$(B-V)_0 \approx 0.5$	at $\varphi_{orb} \approx 0.964$
$(J-K_s)_0 \approx 1.2$	epoch: HJD 2450963.8408 ($\varphi_{orb} \approx 0.88$) (shortly before the rise in mean mag.)

Equation (4), the uncertainty arising mostly from the pulsations) but might be due to the beating of multiple pulsation modes; however, similar phenomena have been reported in the literature for symbiotic binaries (Skopal 2008, for example). Thus, we feel justified in calling attention to this phenomenon, which will be further discussed in section 4. A phase plot of all eclipses which also includes the 1998 fading is shown in Figure 6.

From the considerable increase in eclipse amplitude after the system's rise in mean magnitude, it becomes obvious that the observed brightening is restricted to the primary star of the system. The contribution of the secondary star, whose light dominates at longer wavelengths, remains practically constant, as can be inferred from the EROS-2 I_C measurements (Figure 1).

Obviously, the contribution of the secondary star to the total flux of the system at visual wavelengths is negligible after the considerable brightening. This is further substantiated by the absence of a secondary minimum and an analysis of recent APASS multi-color photometry, which indicates that the system, as expected, gets redder during eclipses, and the amplitude of the eclipses decreases towards longer wavelengths.

The (B–V) index at maximum light is around 1.0 magnitude, which—after correcting for line-of-sight extinction (section 2.1)—corresponds to $(B-V)_0 \approx 0.2$ mag., getting progressively redder as eclipse sets in. Although there are no observations at mid-eclipse, APASS measurements at a phase of $\varphi \approx 0.964$, which is equivalent to 36.414 days before mid-eclipse, indicate a B–V index of 1.24 mag. (approximating to $(B-V)_0 \approx 0.5$ mag.) and a decrease of 2.6 mag. in V and 1.15 mag. in the Sloan i' band relative to the maximum magnitude. Additional observations at phase $\varphi \approx 0.993$ (7.081 days before mid-eclipse) show the star below the detection limit in the B and V bands while the observed decrease in brightness relative to the measurements at phase $\varphi \approx 0.964$ are only 1.3 mag. and 0.57 mag. in the g' and r' bands, respectively, which confirms the shrinking of the amplitude towards the red end of the spectrum. This is illustrated in Figure 7, which shows a detailed view of the primary eclipse based on BVg'r'i' photometry from APASS and V data from ASAS-3. These findings are also confirmed by EROS-2 data; the amplitude of the 2001 eclipse is 0.2 mag. (I_C) and 0.9 mag. (V), respectively.

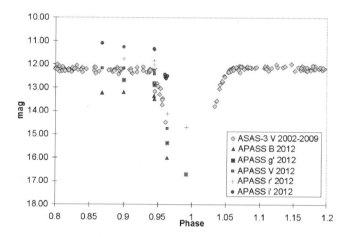

Figure 7. Close-up view of the primary eclipse. The plot is based on ASAS-3 V and APASS multi-color photometry, as indicated in the legend on the right side. The data have been folded with the ephemeris given in Equation (4). Note the decreasing amplitude of the eclipse towards longer wavelengths.

3.2. Spectral energy distribution

Recent photometric measurements from various online catalogue sources are presented in Table 4. Because of the significant line-of-sight dust extinction to ASAS J174600-2321.3 ($A_V \approx 2.4$ mag. and $E(B-V) \approx 0.78$ mag.; Schlafly and Finkbeiner (2011); section 2.1), extinction-corrected values are given for the wavelength range 0.43 μm (Johnson B) < λ < 7.68 μm (GLIMPSE [8.0]).

The spectral energy distribution (SED) of ASAS J174600-2321.3, based on the extinction-corrected values, is shown in Figure 8. APASS data in all five passbands were available from two epochs; APASS data taken at maximum visual light are denoted by triangles, APASS data taken near mid-eclipse at an orbital phase of $\varphi \approx 0.964$ (italicized in Table 4), which corresponds to 36.414 days before mid-eclipse, are denoted by squares.

According to our expectations, the hot component, which—anticipating our results—is likely a white dwarf (WD), dominates the system's emission at maximum visual light, contributing strongly to the short wavelength region and being thus responsible for the observed rise in visual brightness. The red giant, on the other hand, dominates at longer wavelengths, peaking in flux at about 1.65 μm. Near mid-eclipse (orbital phase

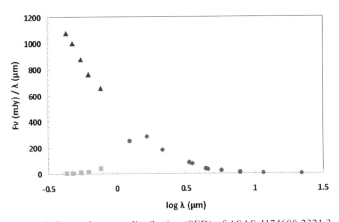

Figure 8. Spectral energy distribution (SED) of ASAS J174600-2321.3 from optical (Johnson B) to mid-infrared (WISE [22]). APASS data taken at maximum visual light are denoted by triangles, APASS data taken near mid-eclipse at an orbital phase of φ ≈ 0.964 are denoted by squares.

φ ≈ 0.964), however, the continuum bluewards of λ ~ 1 μm drops off dramatically as the WD is eclipsed by the red giant star, which dominates the system's optical brightness during this part of the orbital cycle. This confirms the simultaneous presence of two different temperature regimes in ASAS J174600-2321.3, which is typical for symbiotic stars (Allen 1984, for example).

3.3. Spectroscopic characteristics and the types of the primary and secondary stars

The most striking features of the 2007 spectrum (shown in Figure 9) are the sodium (Na I) doublet lines at 5889 Å and 5895 Å, and H-alpha (Hα) absorption at 6563 Å. There is also structure due to titanium oxide (TiO) around 6300 Å, 7200 Å, as well as vanadium oxide (VO) absorption at around 7450 Å. The spectrum in this wavelength region resembles that of an M-type giant; consequently, ASAS J174600-2321.3 was classified as spectral type M0 and rejected as an RCB candidate by Tisserand et al. (2008).

The increased wavelength coverage (3400 Å < λ < 9600 Å) of the 2012 spectrum (shown in Figures 10 and 11), which includes the Balmer and Paschen lines of hydrogen, shed new light on the star. Based on the 2012 spectrum, which is in agreement with the general characteristics of the 2007 spectrum, the star was re-classified as a highly reddened supergiant of spectral type F0 (Tisserand et al. 2013). Reinvestigating the spectrum, we confirm this classification, which is in agreement with the de-reddened color index at maximum visual light of $(B-V)_0 \approx 0.2$ mag. (section 2.2). It is interesting to note that, in contrast to the 2007 spectrum, the 2012 spectrum displays Hα in emission.

Additionally, as in the 2007 spectrum, there is clearly structure due to TiO absorption around 6300 Å, 7200 Å, 7800 Å, and 8300 Å; VO features are present around 7450 Å and, possibly, 7900 Å (Figure 10). The spectrum is obviously composite, the observed TiO and VO absorption being due to the presence of the cool giant, which agrees well with the observed binarity of the system. Judging from the presence of VO bands, which set in at around M7 and increase in strength with declining temperature (Gray and Corbally 2009, for example), the observed $(V-I_C)_0$ index at minimum light (section 2.2),

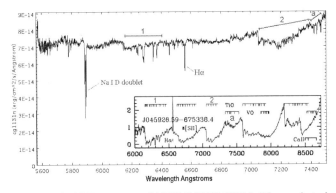

Figure 9. The 2007 spectrum of ASAS J174600-2321.3 (Tisserand et al. 2008). The spectrum was taken outside of an eclipse on HJD ~2454213.9218 and encompasses a wavelength range of 5550 Å < λ < 7500 Å. Note the structure due to TiO (identified by numbers) and VO (identified by letters). In order to facilitate discrimination, the spectrum of a late M-type star (2MASS J04592660–6753383) is shown for comparison (inset), which has been adapted from Wood et al. (2013; ©2013 The Authors. By permission of Oxford University Press on behalf of the Royal Astronomical Society. All rights reserved. For permissions, please email: journals.permissions@oup.com).

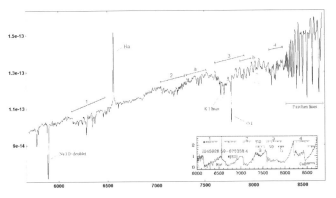

Figure 10. Segment of the 2012 spectrum of ASAS J174600-2321.3 (5700 Å < λ < 8700 Å; Tisserand et al. 2013). The spectrum was taken outside of an eclipse on HJD ~2455762.9661. Note the structure due to TiO (identified by numbers) and VO (identified by letters) and the change from absorption (2007 spectrum) to emission (2012 spectrum) in the Hα line at 6563 Å. In order to facilitate discrimination, the spectrum of a late M-type star (2MASS J04592660-6753383) is shown for comparison (inset), which has been adapted from Wood et al. (2013; ©2013 The Authors. By permission of Oxford University Press on behalf of the Royal Astronomical Society. All rights reserved. For permissions, please email: journals.permissions@oup.com.). Note the CCD fringing that affects the reddest part of the spectrum. The Paschen lines, however, emerge well above the fringing level for a confident identification.

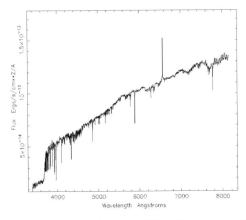

Figure 11. The 2012 spectrum of ASAS J174600-2321.3 to about 8200 Å. The spectrum was taken by Tisserand et al. (2013) outside of an eclipse on HJD ~2455762.9661 and encompasses a wavelength range of 3400 Å < λ < 9600 Å.

and the SED (section 3.2), we conclude that the secondary component of ASAS J174600-2321.3 is a late-M giant (~M7), which perfectly fits all observed parameters.

3.4. Pulsation study

In addition to the mean brightness changes, ASAS J174600-2321.3 also shows semiregular pulsations, which are most obvious in EROS-2 I_C data, for example in and around the 2001 eclipse at around HJD 2452100 (Figure 12). It is obvious that the pulsations continue during the total phase of the observed eclipse, which is proof that the pulsations arise in the late M-type secondary component, as would be expected. Although rather ill-defined throughout most of the covered timespan, there is possible pulsational activity with a timescale around 50 to 60 days in EROS-2 data.

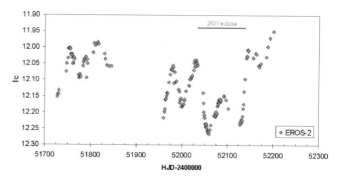

Figure 12. Semiregular pulsations of ASAS J174600-2321.3, based on EROS-2 data (2451726 < HJD < 2452200). EROS-2 R_E magnitudes have been transformed to Cousins I_C (see text for details). Note the continuance of the pulsations during the 2001 eclipse at around HJD 2452100.

After the significant rise in mean brightness related to the primary star, the oscillations are much less clear but still readily visible during maximum light in the ASAS-3 V light curve. However, a period of about 80 to 90 days is now prevalent, especially around HJD 2453500 (Figure 13). While there is no doubt that the pulsations arose in the secondary star before the observed brightening, it is open to debate where they stem from after this event. As the expected contribution of the secondary star at visual wavelengths will now be negligible, the observed pulsations after the brightening are likely to arise in the F-type primary star, which might explain the apparent change in period; however, more data are needed to resolve this matter.

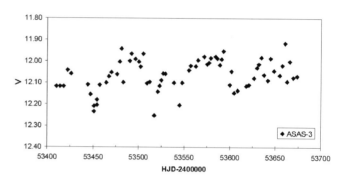

Figure 13. Semiregular pulsations of ASAS J174600-2321.3 after the dramatic brightening, based on ASAS-3 data (2453409 < HJD < 2453673).

As will be shown in the discussions in section 4, the primary star is likely a WD currently in outburst. In this respect, it is interesting to mention that—according to theoretical considerations—the outer portion of the white dwarf envelope might become pulsationally unstable during the first few months after visual maximum has been reached in a symbiotic nova eruption (Kenyon and Truran 1983). Pulsations of comparable period and amplitude have, for example, been observed during the maximum of the symbiotic nova PU Vul (P ≈ 80 d; ΔV ≈ 0.15 mag.; see Kenyon 1986 and references therein).

An analysis of EROS-2 I_C data with different period search algorithms (CLEANest (Foster 1995) and ANOVA (Schwarzenberg-Czerny 1996), as implemented in PERANSO (Vanmunster 2007); PERIOD04 (Lenz and Breger 2005)) suggests a dominant signal at P ≈ 56 days (Figure 14). Although the amplitude of the signal is rather weak and the error margin is considerable due to the semiregular nature of the variability, we adopt this value as a starting point for the computation of distance and other stellar parameters in the following section.

3.5. Distances
3.5.1. Distance in a high extinction area

We have estimated the distance to ASAS J174600-2321.3 using the period-luminosity (P-L) relations for M-type semiregular variables based on Hipparcos parallaxes proposed by Yeşilyaprak and Aslan (2004). Assuming that the M giant dominates the system's light before the outburst in V and, especially, in I_C, we have derived mean magnitudes of ASAS J174600-2321.3 in both passbands from converted EROS-2 observations before the onset of the brightening at around HJD 2451240 (V = 16.3 mag.; I_C = 12.2 mag.). We have used the

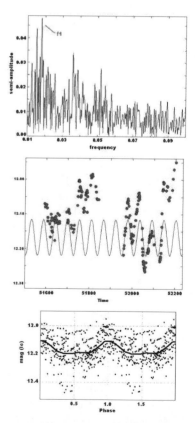

Figure 14. PERIOD04 Fourier graph and period fit (upper panels), based on an analysis of EROS-2 I_C data. The corresponding phase plot (lower panel, with fit curve) is folded with a period of P = 56 days. EROS-2 R_E magnitudes have been transformed to Cousins I (see section 2.1 for details). Note the changes in mean magnitude which are evident from the period fit and phase plot.

corresponding P-L relations for the Johnson V and Cousins I bands (Yeşilyaprak and Aslan 2004; especially their Table 2), adopting a period value of P = 56 days (section 3.4):

$$M_V = 2.89 (\pm 0.49) \times \log(P) - 5.30 (\pm 0.85) \quad (5)$$

$$M_{Ic} = 0.83 (\pm 0.48) \times \log(P) - 4.79 (\pm 0.85) \quad (6)$$

From this, we derive absolute magnitudes of $M_V \approx -0.25$ mag. and $M_{Ic} \approx -3.35$ mag. for the red giant component of ASAS J174600-2321.3. Taking into account the line-of-sight extinction $A_V \approx 2.4$ mag. and $A_I \approx 1.4$ mag., calculated using the estimates of Schlafly and Finkbeiner (2011), we derive distances of ~6.7 kpc and ~6.8 kpc from Equations (5) and (6), respectively. Similar results are obtained using the Hipparcos K-band P-L diagram for semiregular variables from Bedding and Zijlstra (1998), which further substantiates our estimates.

Considering the sources of error (intrinsic uncertainties of the P-L relations, conversion of EROS data to standard passbands, semiregularity of the period), it becomes obvious that our distance estimate is only a first approximation that further studies may build upon in the future. However, we feel justified in concluding that ASAS J174600-2321.3 (Galactic coordinates l, b = 2.829°, 4.814°) is situated in the Galactic Bulge, at a distance of about 2 kpc from the Galactic Center.

3.5.2. M giant radius, mass estimates, and orbital distance

We have assessed the effective temperature of the red giant from the proposed spectral class of ~M7 (section 3.3) and derive $T_{eff} \approx 3,150$ K. From period and amplitude of the observed pulsations, we have assessed the luminosity of the giant at $\log(L_\star / L_\odot) \approx 3.64$ using Figure 6.2 of Fraser (2008).

By comparing these results to the evolutionary tracks for AGB stars of about solar metallicity (Marigo et al. 2008 and Girardi et al. 2010; http://stev.oapd.inaf.it/cgi-bin/cmd), we derive a mass of $M_g \approx 1.5\, M_\odot$ for the red giant star (Figure 15), which is in good agreement with the mass estimates for symbiotic binaries compiled from the literature by Mikołajewska (2003, 2010). Based on the aformentioned sources, we furthermore adopt a mass of $M_{wd} \approx 0.5\, M_\odot$ for the white dwarf.

Assuming that the red giant star pulsates in the first overtone, as is observed and proposed for many semiregular variables of similar period (for example, Percy and Parkes 1998; Soszyński et al. 2013), we have estimated the radius of the red giant using Equation (4) of Mondal and Chandrasekhar (2005):

$$\log P = 1.59 \log(R / R_\odot) - 0.51 \log(M / M_\odot) - 1.60 \quad (7)$$

Using P = 56 days and $M_g = 1.5\, M_\odot$, we derive a radius of $R_g \approx 145\, R_\odot$, which is in accordance with the values gleaned from the "Catalogue of Apparent Diameters and Absolute Radii of Stars" (Pasinetti Fracassini et al. 2001) for other stars of similar spectral type and period.

Using the third Keplerian law and assuming a circular orbit, we have estimated the distance between the red giant and the

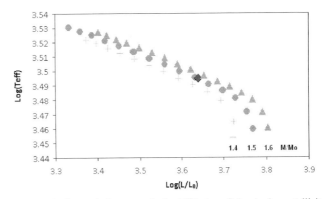

Figure 15. Stellar evolutionary tracks for AGB stars of about solar metallicity and different masses, adapted from Marigo et al. (2008) and Girardi et al. (2010) and accessed through http://stev.oapd.inaf.it/cgi-bin/cmd. The position of the red giant component of ASAS J174600-2321.3 is marked by the diamond.

WD to be $R \approx 2.5$ AU. This results in an essentially consistent picture that is in general accordance with the data compiled by Mikołajewska (2003) for other systems. Furthermore, the adopted configuration is able to reproduce the observed eclipse duration of about 111 days (section 3.1).

We used a simplified model, therefore the given values have to be treated with caution and serve only as a first approximation. For example, it is possible that the radius of the red giant star has been underestimated because of the unknown exact orbital inclination of the system. Only detailed studies will be able to shed more light on the system's parameters.

4. Discussion

4.1. General discussion

Taking into account all evidence, we conclude that ASAS J174600-2321.3 consists of a primary star that exhibits the spectral features of an early F-type supergiant in outburst and a secondary component of spectral type late M (~M7). The primary star, as will be shown in the following discussions, is a hot, compact object – probably a WD. Considering the orbital period, the observed time of totality, its significant contribution to the total flux of the system at longer wavelengths, and its intrinsic variability, we deduce that the cool component is an evolved giant.

Judging from the light curve peculiarities and the observed spectroscopic features (notably the Hα emission in the 2012 spectrum), it seems likely that both components interact. Circumstellar (or circumbinary) gas and dust might be involved, either in a disk or shell. Following this scenario, the observed long-term rise in brightness might be due to the dispersion of an optically thick circumstellar shell of gas and dust—a phenomenon regularly seen in young or evolved stars of high luminosity. However, this is not supported by near- and mid-infrared measurements. Furthermore, no bright state has been recorded during the past, although this evidence is based on only a few scattered measurements before the onset of the EROS project (section 2.2). Nevertheless, it seems likely that the observed brightening does not constitute the recovery from a faint state but rather an outburst. Obviously, then, being the seat of the brightening, the primary star has developed the observed F-type supergiant features in outburst.

Symbiotic stars (GCVS-type ZAND; Samus *et al.* 2007–2013) and symbiotic novae (also called "very slow novae"; GCVS-type NC) are subject to prolonged outbursts and associated with long-period binary systems. The symbiotic star/nova scenarios are explored in more detail in the following sections.

4.2. The symbiotic star scenario

The term symbiotic star was coined by Merril in 1941 and applied to stars whose spectra are characterized by a combination of the features of a hot source and a cool continuum with prominent absorption features reminiscent of a late-type star (see for example, Sahade 1982; Crocker *et al.* 2001). In fact, symbiotic stars are long-period interacting binary systems, comprising a compact, hot object as accretor (usually a white dwarf), and a yellow or red giant star as donor (for example, Kenyon 1986; Skopal 1998). These systems are characterized by complex photometric variability, combining eclipses, pulsations, and outbursts on diverse scales, which is akin to what is seen in ASAS J174600-2321.3.

As has been pointed out above, ASAS J174600-2321.3 exhibits a composite spectrum characterized by the features of an early F-type supergiant, which might originate from the pseudophotosphere of a WD in eruption, and the molecular absorption lines of a late M star. It also exhibits photometric variability characteristic of symbiotic stars, which will be discussed below. Several characteristic light curve features are indicated in Figure 16 and marked by the corresponding lower case letters used in the explanatory text below.

(a) During quiet phases, the light curves of symbiotic stars are characterized by wave-like orbitally related variations (for example, Skopal 2008; Skopal *et al.* 2012; their Figures 15 and 16). During quiescence, radiation from the hot component (partly) ionizes the wind from the late-type giant and the neutral circumbinary material, giving rise to a local H_{II} region, whose nebular radiation—according to Skopal (2008)—is the source of the orbitally related variations. Other phenomena, like ellipsoidal variations, might be involved (Gromadzki *et al.* 2013). In amplitude and shape, those variations are reminiscent of the "arc" seen in the light curve of the present object between HJD ~2451200 and HJD ~2452100, before the onset of activity in the system.

(b) In symbiotic systems with high orbital inclination, significant changes in the minima profiles are observed between quiescence and activity. The local H_{II} region has an asymmetrical shape due to orbital motion and interaction with the stellar winds. The nebular emission varies as a function of orbital phase, producing broad, flat minima during quiescence, which occur prior to the time of spectroscopic conjunction at an orbital phase of $\varphi \approx 0.9$ (for a more detailed discussion, see Skopal 1998 and Skopal 2008). This leads to systematic changes in the O–C residuals of the minima times in eclipsing symbiotic binaries (for example, Skopal 2008, especially section 6). As has been described in section 3.1, the "V-shaped" 1998 fading occurred at an orbital phase of $\varphi \approx 0.9$. It is comparable in amplitude and shape to what has been described for other symbiotic binaries (Skopal 2008). It is thus intriguing to interpret the 1998 fading in this vein; however, as has been

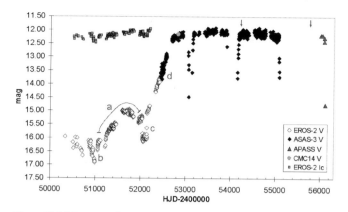

Figure 16. Light curve of ASAS J174600-2321.3, based on EROS-2, ASAS-3 and APASS V data. Symbols and data sources are the same as in Figure 1. The epochs of the spectra are indicated by arrows (see section 3.3 for details). Several light curve features, discussed in the text, are marked by lowercase letters.

pointed out above, it may simply be the result of the interference of different pulsation modes. Only long-term light curve studies will be able to shed more light on this issue.

(c) With the onset of activity in the system, the optical region usually becomes dominated by the radiation from the pseudophotosphere of the compact object and a narrow, "Algol-like" eclipse is observed at the inferior conjunction of the giant (Skopal 2008). This is exactly what has been observed in the present object after the onset of activity that led to the dramatic brightening of the primary component.

(d) Several types of brightenings are observed in symbiotic binaries (for example, Leibowitz and Formiggini 2006; Skopal *et al.* 2012), encompassing slow, symbiotic-nova-type outbursts, short-term flares, and ZAND-type outbursts of medium amplitude (~3 mag. in V). Active phases can last for many years, which also seems to be the case with ASAS J174600-2321.3. However, the observed outburst differs from that of classical symbiotic stars. The amplitude is larger, and the hot star comes to completely dominate the optical emission; superposed irregular variations are not seen and the outburst is very slow and much quieter than in other sources, which is reminiscent of the behavior of symbiotic novae.

4.3. An eclipsing, symbiotic nova?

Symbiotic novae are a small class of objects that have been variously classified as either classical novae (for example, RT Ser, RR Tel) or symbiotic stars (for example, V1016 Cyg, HM Sge) in the literature (Kenyon and Truran 1983). They are a rare subclass of symbiotic stars (e.g. Kenyon 1986) and distinguished from the classical specimens by their outburst activity (occasional eruptions on timescales from months to years and amplitudes from 1 to 3 mag. for ZAND systems; a single outburst of up to ~7 mag. lasting for decades for NC systems; see, for example, Kenyon and Truran 1983).

Symbiotic novae are themselves often subdivided into two groups: very slow novae (GCVS-type NC) or very fast recurrent novae (GCVS-type NR) with recurrence times of the order of several years (Kato 2002), the deciding criterion being the mass of the WD. According to the aforementioned author, the term symbiotic nova is usually reserved for the former class of objects, which we adhere to throughout this paper; the latter group is of no interest in the analysis of the present object.

Symbiotic novae eruptions develop very slowly; the systems may remain at maximum light for decades, after which they fade very slowly. Apart from that, symbiotic novae share common properties with classical symbiotic stars: they are always long-period ($P_{orb} \geq 800$ days; Mikołajewska 2010), interacting binary stars comprising an M giant. In fact, there exists no clear cut boundary between classical symbiotic stars and symbiotic novae (Skopal *et al.* 2012, for example).

As the underlying mechanism in symbiotic novae eruptions, Kenyon and Truran (1983) propose hydrogen shell flashes involving low-luminosity WDs accreting matter at low rates. Flashes in a degenerate WD envelope ("degenerate flash") lead to evolution from a relatively faint object to a luminous A-F supergiant at maximum visual light, which is in agreement with the spectroscopic characteristics of the present object observed at maximum visual brightness. Systems with WDs undergoing degenerate flashes (as is proposed for RR Tel and RT Ser) remain at visual maximum for decades and lose mass in a low velocity wind (Kenyon and Truran 1983). The higher excitation emission lines, characteristic of most symbiotic systems, fade during the rise in visual brightness (Kenyon and Truran 1983; Corradi *et al.* 2010) and return during the decline, indicating that the WD increases again in effective temperature.

The above mentioned findings are in agreement with the system configuration and spectral characteristics of ASAS J174600-2321.3. No high excitation emission lines are present in our spectra and there is no clear evidence for a rapidly expanding envelope, except for the change in the Hα line from absorption in the 2007 spectrum to emission in the 2012 spectrum, which might be interpreted along the lines of mass loss in a low velocity wind.

Furthermore, as has been pointed out in section 4.2, the general characteristics of the observed rise in brightness in the present object are similar in respect to amplitude and maximum peak shape to what has been observed for many symbiotic novae, see, for example, the historical light curve of RR Tel shown in Figure 17 or the optical light curve of V1016 Cyg in Kenyon (1986; Figure A.11). According to ASAS-3 and APASS data, the system has remained at visual maximum light for more than ~3,200 days. Furthermore, the star is completely saturated in OGLE-III and OGLE-IV data, which reaches up to the present time (Soszyński, 2014; private communication). Additionally, we have started a long-term photometric monitoring program on the star, the results of which will be presented in an upcoming paper. First results haven shown that, in the timespan from HJD 2456855.6 to HJD 2456883.5, the star's brightness fluctuated around a mean magnitude of 12.27 mag (V), which is strong evidence that the outburst has already continued for more than ~4,100 days and still continues.

In summary, an accretion scenario involving the hot primary star of ASAS J174600-2321.3 as accretor and the semiregular, late-type secondary component as donor seems to be the most promising scenario which is capable of explaining all observed light curve features. Thus, we feel confident that the star is an eclipsing symbiotic system. Because of the peculiarities of the observed outburst (amplitude, duration, shape),

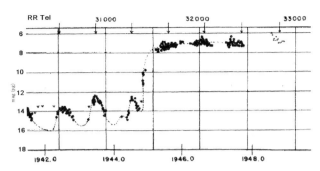

Figure 17. Historical light curve of RR Tel, adapted from Mayall (1949; data prior to 1948 are photographic from Harvard College Observatory plates; 1948 data are visual observations from the AAVSO).

we propose ASAS J174600-2321.3 as a symbiotic nova of type NC. However, further photometric and spectroscopic data are needed to confirm this.

5. Conclusion

By an analysis of all available data from the literature, public sky surveys, and photometric catalogues, we have identified the star ASAS J174600-2321.3 as a long-period eclipsing binary with an orbital period of $P_{orb} = 1,011.5$ days. The primary star, which is likely a white dwarf in outburst, exhibits spectral characteristics of a reddened, early F-type supergiant at visual maximum light; the secondary star is a giant of spectral type late M (~M7). The system underwent a significant increase in brightness (~4 mag (V)) during the recent past, which is related to the primary component. We have discussed the origin of the observed brightening and favor the scenario of an eclipsing, symbiotic binary, which is capable of explaining the observed, complex light curve features. Because of the peculiarities of the observed outburst (large amplitude, duration, slow development) and spectral characteristics, we propose ASAS J174600-2321.3 as a symbiotic nova of type NC. Magnitude range, periods, stellar parameters, and binary separation of ASAS J174600-2321.3, as derived in the present paper, are presented in Table 5.

The outburst has already lasted for more than ~4,100 days and seems to continue to the present day. We have started a long-term photometric monitoring of the system, the results of which will be presented in an upcoming paper.

6. Acknowledgements

This research has made use of the SIMBAD and VizieR databases operated at the Centre de Données Astronomiques (Strasbourg) in France. This work has also made use of EROS-2 data, which were kindly provided by the EROS collaboration. The EROS (Expérience pour la Recherche d'Objets Sombres) project was funded by the CEA and the IN2P3 and INSU CNRS institutes. Furthermore, this research has employed data products from the Two Micron All Sky Survey, which is a joint project of the University of Massachusetts and the Infrared Processing and Analysis Center/California Institute of Technology, funded by the National Aeronautics and Space Administration and the National Science Foundation, and the

Table 5. Magnitude range, periods, stellar parameters, and binary separation of ASAS J174600-2321.3, as derived in the present paper.

Parameter	Value
Magnitude Range (V)	11.9–16.9 mag.
Orbital Period	1,011.5 days
Stellar Parameters	
(Red Giant)	$M_g \approx 1.5\ M_\odot$
	$R_g \approx 145\ R_\odot$
	$T_{eff}^g \approx 3{,}130$ K
	Spectral type ~M7
(White Dwarf)	$M_{wd} \approx 0.5\ M_\odot$
Binary Separation	$R \approx 2.5$ AU

Wide-field Infrared Survey Explorer, which is a joint project of the University of California, Los Angeles, and the Jet Propulsion Laboratory/California Institute of Technology, funded by the National Aeronautics and Space Administration. This research has also made use of the NASA/IPAC Infrared Science Archive, which is operated by the Jet Propulsion Laboratory/California Institute of Technology, under contract with the National Aeronautics and Space Administration. The authors would like to thank Franz-Josef (Josch) Hambsch, Belgium, for acquiring photometric observations of our target, and the anonymous referee for helpful comments and suggestions that helped to improve the paper.

References

Adelman-McCarthy, J. K., et al. 2011, *The SDSS Photometric Catalog*, Release 8, VizieR On-line Data Catalog: II/306

Allen, D. A. 1984, *Astrophys. Space Sci.*, **99**, 101.

Bedding, T. R., and Zijlstra, A. A. 1998, *Astrophys. J.*, **506**, L47.

Corradi, R. L. M., et al. 2010, *Astron. Astrophys.*, **509**, 9.

Crocker, M. M., Davis, R. J., Eyres, S. P. S., Bode, M. F., Taylor, A. R., Skopal, A., and Kenny, H. T. 2001, *Mon. Not. Roy. Astron. Soc.*, **326**, 781.

Dopita, M., Hart, J., McGregor, P., Oates, P., Bloxham, G., and Jones, D. 2007, *Astrophys. Space Sci.*, **310**, 255.

Ducati, J. R., Bevilacqua, C. M., Rembold, S. B., and Ribeiro, D. 2001, *Astrophys. J.*, **558**, 309.

Evans, D. W., Irwin, M. J., and Helmer, L. 2002, *Astron. Astrophys.*, **395**, 347.

Fitzpatrick, E. L. 1999, *Publ. Astron. Soc. Pacific*, **111**, 63.

Foster, G. 1995, *Astron. J.*, **109**, 1889.

Fraser, O. J. 2008, *Properties of long-period variables from the MACHO Project*, Proquest Dissertations and Theses 2008, Section 0250, Part 0606 [Ph.D. dissertation], University of Washington, publication number AAT 3328397, source DAI-B 69/09, Mar 2009.

Girardi, L., et al. 2010, *Astrophys. J.*, **724**, 1030.

Gray, R. O., and Corbally, C. J. 2009, *Stellar Spectral Classification*, Princeton Univ. Press, Princeton, NJ.

Gromadzki, M., Mikołajewska, J., and Soszyński, I. 2013, *Acta Astron.*, **63**, 405.

Henden, A. A., et al. 2012, AAVSO Photometric All-Sky Survey, data release 3 (http://www.aavso.org/apass).

Jester, S., et al. 2005, *Astron. J.*, **130**, 873.

Kato, M. 2002, in *The Physics of Cataclysmic Variables and Related Objects*, ASP Conf. Proc. 261, eds. B. T. Gänsicke, K. Beuermann, and K. Reinsch, Astronomical Society of the Pacific, San Francisco, 595.

Kenyon, S. J. 1986, *The Symbiotic Stars*, Cambridge Univ. Press, Cambridge.

Kenyon, S. J., and Truran, J. W. 1983, *Astrophys. J.*, **273**, 280.

Leibowitz, E. M., and Formiggini, L. 2006, *Mon. Not. Roy. Astron. Soc.*, **366**, 675.

Lenz, P., and Breger, M. 2005, *Commun. Asteroseismology*, **146**, 53.

Marigo, P., Girardi, L., Bressan, A., Groenewegen, M. A. T., Silva, L., and Granato, G. L. 2008, *Astron. Astrophys.*, **482**, 883.

Mayall, M. W. 1949, *Bull. Harvard Coll. Obs.*, No. 919, 15.

Mikołajewska, J. 2003, in *Symbiotic Stars Probing Stellar Evolution*, eds. R. L. M. Corradi, J. Mikołajewska, and T. J. Mahoney, Astron. Soc. Pacific Conf. Proc. 303, Astronomical Society of the Pacific, San Francisco, 9.

Mikołajewska, J. 2010, in *The Proceedings of the Physics of Accreting Compact Binaries* (arXiv:1011.5657).

Mondal, S., and Chandrasekhar, T. 2005, *Astron. J.*, **130**, 842.

Pasinetti Fracassini, L. E., Pastori, L., Covino, S., and Pozzi, A. 2001, *Astron. Astrophys.*, **367**, 521.

Percy, J. R., and Parkes, M. 1998, *Publ. Astron. Soc. Pacific*, **110**, 1431.

Pojmański, G. 2002, *Acta Astron.*, **52**, 397.

Pojmański, G., Pilecki, B., and Szczygiel, D. 2005, *Acta Astron.*, **55**, 275.

Renault, C., et al. 1998, *Astron. Astrophys.*, **329**, 522.

Rodgers, A. W., Conroy, P., and Bloxham, G. 1988, *Publ. Astron. Soc. Pacific*, **100**, 626.

Sahade, J. 1982, in *The Nature of Symbiotic Stars*, D. Reidel Publishing Co., Dordrecht, 1.

Samus, N. N., et al. 2007–2013, *General Catalogue of Variable Stars*, VizieR On-line Data Catalog (http://cdsarc.u-strasbg.fr/viz-bin/Cat?B/gcvs).

Schlafly, E. F., and Finkbeiner, D. P. 2011, *Astrophys. J.*, **737**, 103.

Schwarzenberg-Czerny, A. 1996, *Astrophys. J.*, **460**, 107.

Skopal, A. 1998, *Astron. Astrophys.*, **338**, 599.

Skopal, A. 2008, *J. Amer. Assoc. Var. Star Obs.*, **36**, 9.

Skopal, A., Shugarov, S., Vanko, M., Dubovský, P., Peneva, S. P., Semkov, E., and Wolf, M. 2012, *Astron. Nachr.*, **333**, 242.

Soszyński, I. 2014, private communication.

Soszyński, I., Wood, P. R., and Udalski, A. 2013, *Astrophys. J.*, **779**, 167.

Tisserand, P., et al. 2007, *Astron. Astrophys.*, **469**, 387.

Tisserand, P., et al. 2008, *Astron. Astrophys.*, **481**, 673.

Tisserand, P., Clayton, G. C., Welch, D. L., Pilecki, B., Wyrzykowski, L., and Kilkenny, D. 2013, *Astron. Astrophys.*, **551A**, 77.

Vanmunster, T. 2007, Light Curve and Period Analysis Software, PERANSO (http://www.peranso.com/).

Wood, P. R., Kamath, D., and Van Winckel, H. 2013, *Mon. Not. Roy. Astron. Soc.*, **435**, 355.

Yeşilyaprak, C., and Aslan, Z. 2004, *Mon. Not. Roy. Astron. Soc.*, **355**, 601.

Zacharias, N., Finch, C. T., Girard, T. M., Henden, A., Bartlett, J. L., Monet, D. G., and Zacharias, M. I. 2012, *The Fourth U.S. Naval Observatory CCD Astrograph Catalog* (UCAC4), VizieR On-line Data Catalog (http://cdsarc.u-strasbg.fr/viz-bin/Cat?I/322).

New Variable Stars Discovered by the APACHE Survey. II. Results After the Second Observing Season

Mario Damasso
INAF-Astrophysical Observatory of Torino, Via Osservatorio 20, I-10025 Pino Torinese, Italy; Astronomical Observatory of the Autonomous Region of the Aosta Valley, fraz. Lignan 39, 11020 Nus (Aosta), Italy; damasso@oato.inaf.it and m.damasso@gmail.com

Lorenzo Gioannini
Department of Physics, University of Trieste, Via Tiepolo 11, I-34143 Trieste, Italy

Andrea Bernagozzi
Enzo Bertolini
Paolo Calcidese
Albino Carbognani
Davide Cenadelli
Astronomical Observatory of the Autonomous Region of the Aosta Valley, fraz. Lignan 39, 11020 Nus (Aosta), Italy

Jean Marc Christille
Department of Physics, University of Perugia, Via A. Pascoli, 06123 Perugia, Italy; Astronomical Observatory of the Autonomous Region of the Aosta Valley, fraz. Lignan 39, 11020 Nus (Aosta), Italy

Paolo Giacobbe
Luciano Lanteri
Mario G. Lattanzi
Richard Smart
Allesandro Sozzetti
INAF-Astrophysical Observatory of Torino, Via Osservatorio 20, I-10025 Pino Torinese, Italy

Received January 19, 2015; revised February 6, 2015; accepted February 11, 2015

Abstract Routinely operating since July 2012, the APACHE survey has celebrated its second birthday. While the main goal of the project is the detection of transiting planets around a large sample of bright, nearby M dwarfs in the northern hemisphere, the APACHE large photometric database, consisting of hundreds of different fields, represents a relevant resource to search for and provide a first characterization of new variable stars. We celebrate here the conclusion of the second year of observations by reporting the discovery of 14 new variables.

1. Introduction

APACHE (A PAthway to the Characterization of Habitable Earths) is a ground-based photometric survey specifically designed to search for transiting planets orbiting bright, nearby early-to-mid M dwarfs (Sozzetti *et al.* 2013), and it is mainly sensitive to companions with orbital periods up to 5 days. The project is based at the Astronomical Observatory of the Autonomous Region of the Aosta Valley (OAVdA), located in the Western Italian Alps, and the scientific observations started in July 2012. The survey utilizes an array of five automated 40-cm telescopes to monitor hundreds of M-dwarfs and, together with the search for transit-like signals in their light curves, we also look for new variables among the stars that fall in the fields of view of the telescopes, each centered on the target cool stars. We presented in Damasso *et al.* (2014) (hereafter Paper I) more than 80 new variable stars that we discovered after the first year of the survey. Here we announce a list of 14 new variables, not appearing in the AAVSO International Variable Star Index (VSX), that we have detected by the end of the second season of APACHE. The number of findings appears to be much less than those discussed in Paper I because of the following reasons. First, the number of new fields surveyed between 2013 and 2014 was not as large as that of the M dwarfs observed during the first season. Until the end of September 2013 we observed 257 different fields, and 190 between October 2013 and the end of September 2014. Of them, only 42 are newly observed fields. This large overlap between the targets observed over two consecutive seasons mainly reflects a key requirement of the survey. In fact, from a single observing site the detection of transiting planets with orbital periods up to 5 days naturally requires observations of the same targets which extend to more than one year. Moreover, weather conditions heavily influenced the observations during the second season, and the amount of data necessary for a good characterization of the variables was collected for a relatively low number of fields.

The time series of the APACHE differential magnitudes for the variables discussed here are available upon request.

2. Instrumentation and methods

We refer the reader to Paper I, Christille *et al.* (2013), and Sozzetti *et al.* (2013) for a detailed description of the observation strategy, hardware, and software systems which characterize the APACHE survey that did not undergo any relevant change during the second season. For convenience, here we repeat only the main key parameters of the survey. Each APACHE telescope is characterized by a pixel scale of 1.5 arcsec/pixel and a field of view of 26' × 26'. While four telescopes observe in the Johnson-Cousin Ic filter, one instrument uses a V filter to observe the brightest M dwarfs, in order to increase the exposure times (that in the Ic band would be very short) and reduce the overheads. The APACHE observations are carried out in focus, with exposure times, kept fixed during the sessions independently from the seeing (which typically ranges between 1.2 to 3 arcsec), that are in the range of 3 to 180 seconds and are optimized for the M dwarfs, which are the primary targets. The magnitudes of the target M dwarfs, over which the APACHE observing strategy is tailored, vary within 8–16.5 in the V band and 5.5–13 in the J band. We use a circular observing schedule, with each target being re-pointed typically ~20 to 25 minutes after the last observation. This cadence should be optimal for collecting enough data points which fall in the portion of the light curve showing a transit, if it is in progress, that usually is expected to last for 1 to 3 hours. Each time a target is pointed, three consecutive exposures are taken and usually the average value of the corresponding differential magnitudes is used for light curve analysis.

Light curves of M dwarfs and all the field stars are produced by the pipeline TEEPEE (see Paper I), which we developed mostly in IDL programming language specifically for APACHE. (IDL is the registered trademark of Exelis Visual Information Solutions.) Basically, TEEPEE performs ensemble differential aperture photometry by automatically selecting i) the best set of comparison stars among the brightest in the field, i.e. those with light curves showing the lowest r.m.s, varying in number from field to field, and ii) the best aperture among twelve test aperture radii in the range 3.5 to 9 pixels. The selection is made only once for the primary target M dwarf, and the same set of comparison stars and aperture radius are applied to all the field stars.

As indicated in Paper I, also for the objects discussed here no spectroscopic observations and analysis could be performed to better characterize their astrophysical properties and validate our tentative variability classifications. Their classification is thus based on photometric data only. While the detection of new variables originated from the APACHE database, as done for Paper I we also examined the on-line light curve archive of the SuperWASP survey (http://exoplanetarchive.ipac.caltech.edu/), where data from the first public data release, collected between 2004 and 2008 for nearly 18 million targets, are stored. Where applicable, the SuperWASP data represent a very useful resource for a more accurate variability classification and determination of physical parameters, as for example the periodicity, of new variable stars. Each of the SuperWASP observatories (one based in the island of La Palma, the other located at the site of the South African Astronomical Observatory) consists of eight wide-angle cameras (field of view 61 square degrees) that simultaneously monitor the sky for planetary transit events. (More details about SuperWASP can be found at http://www.superwasp.org/.) The SuperWASP pixel scale is quite high (13.7 arcsec/pixel), and before using the data, we carefully checked if stellar companions were present so close to the variable of interest as to fall within the SuperWASP aperture radius and contaminate the photometry.

Our results and the main information about the variables discussed in this work are listed in Table 1. In discussing some of the targets we used the information about the color excess E(B–V) integrated along the line of sight. We stress here that, without knowing the distance to the star, the color excess corresponding to the star location cannot be directly estimated, and we use the integrated E(B–V) along the line of sight only as a rough indication of how much the dust absorption can influence the correct interpretation of the photometric data. In all the cases the Galactic dust reddening for a line of sight was estimated through the maps provided by the NASA/IPAC Infrared Science Archive (http://irsa.ipac.caltech.edu/applications/DUST/), using the determinations based on the work of Schlafly and Finkbeiner (2011).

As a final note, the finding charts of all the stars are publicly available through the Aladin Sky Atlas service (http://aladin.u-strasbg.fr/aladin.gml).

3. Results for individual variables

3.1. UCAC4 837-000728

We analyzed the light curve with the Lomb-Scargle (L-S; Scargle 1982) algorithm and detected a clear sinusoidal-like modulation with a dominant period of 0.185848 day. The L-S periodogram and the folded light curve are shown in Figure 1. The periodicity and the amplitude of the light curve variations (~0.2 mag. in I band) are compatible with those of a δ Scuti (DSCT) type, for which the highest frequency represents the principal pulsation mode, but the colors of the target are indicative of an early K rather than a A-F type star, if a main-sequence star is assumed. Nonetheless, the color excess E(B–V) integrated along the line of sight is ~0.3 mag., and this suggests that the star could be significantly reddened due to interstellar dust.

We can confidently exclude that this object is an eclipsing binary because, by doubling the best period found, the typical shape of this type of variable does not appear and the original shape is not preserved in any form. Finally we note that two other peak frequencies appear in the periodogram, corresponding to periods of ~0.228271 and 0.156723 day in order of their significance. They could be interpreted as secondary pulsation frequencies.

3.2. UCAC4 612-044588

This variable appears to be a short-period (~21 hours) eclipsing binary showing a distorted light curve and a well-defined primary minimum (Figure 2). Data from the APACHE survey, collected between January 8, 2013, and April 24, 2014, provide convincing evidence of the presence of the secondary minimum shifted of 0.5 in phase with respect to the primary.

Table 1. Main information and results about the new variables discovered by the APACHE survey after the second season.

Star	R.A. (J2000)[a] h	Dec. (J2000)[a] °	V	B–V	V–J	V–K	E(B–V)[b]	Period (days)	Amplitude (mag.)	T_0 (HJD–2455000)	Type	Gal. Lat. °
UCAC4 837-000728	6.8333315	+77.2397420	14.044±0.03	0.78±0.1	1.52±0.04	1.91±0.04	0.283	0.185848±0.000001	~0.2 (APACHE I band)	1367.274897	DSCT	+14.43
UCAC4 612-044588	140.1938968	+32.3455925	13.196±0.03	0.68±0.04	1.33±0.04	1.85±0.04	0.017	0.88965±0.00001	~0.09 (APACHE I band)	1300.683395	EB[c]	+44.4
1SWASPJ092046.54+322044.1									~0.13 (SuperWASP)			
UCAC4 667-058562	209.5897209	+43.3495298	13.03±0.02	0.95±0.04	1.61±0.03	2.34±0.03	0.008	10.62±0.02	~0.07 (APACHE I band) ~0.026 (SuperWASP)	1724.4068	ROT	+68.8
UCAC4 854-011628	270.4706342	+80.6372137	15.13±0.01	0.70±0.04	1.37±0.03	1.76±0.05	0.076	0.30292±0.00001	~0.3 (APACHE V band)	1809.3644	EW	+28.7
UCAC4 609-091606	297.6272386	+31.6943759	13.14±0.01	0.74±0.01	1.64±0.02	2.03±0.02	1.32	1.738±0.001	>0.37 (APACHE I band)	1150.381793	EB	+2.6
UCAC4 610-092815	297.6690953	+31.9639217	15.74±0.01	0.78±0.10	1.51±0.03	1.95±0.05	1.10	0.42706±0.00001	~0.45 (APACHE I band) ~0.42 (APACHE V band)	1150.4765	EW	+2.77
UCAC4 621-119831	319.2999568	+34.1319917	13.57±0.04	1.76±0.06	3.58±0.04	4.76±0.04	0.131	—	>0.2 (APACHE V band)	1829.515146	L	–10.47
UCAC4 620-119316	319.3728765	+33.9577598	11.98±0.04	1.84±0.06	3.54±0.05	4.72±0.04	0.144	—	>0.1 (APACHE V band)	1829.515146	L	–10.63
UCAC4 620-119722	319.7340386	+33.9108506	14.11±0.11	0.43±0.11	0.96±0.11	1.25±0.11	0.15	0.2778±0.0001	~0.15 (APACHE V band) ~0.1 (SuperWASP)	1385.653719	DSCT	–10.88
1SWASPJ211856.22+335439.3												
UCAC4 673-106048	326.4054627	+44.4972139	13.70±0.01	1.92±0.03	5.86±0.02	7.28±0.03	0.306	—	~0.2 (APACHE I band)	1532.607956	LB	–6.76
UCAC4 858-013784	343.6031771	+81.5279153	12.04±0.01	0.53±0.01	1.07±0.02	1.31±0.03	0.238	0.07077±0.00001	~0.02 (APACHE I band)	1385.653719	DSCT	+19.67
UCAC4 849-017521	348.9477371	+79.6295775	12.18±0.03	0.47±0.03	1.08±0.04	1.40±0.04	0.198	0.5988±0.0001	~0.01 (APACHE I band)	1639.405	EW	~17.6
UCAC4 848-018678	349.0945677	+79.4436050	13.06±0.05	1.54±0.08	2.80±0.05	3.72±0.05	0.2	~41	~0.02 (APACHE I band)	1367.2860	ROT? SR?	~17.6
UCAC4 849-017658	350.5989427	+79.7416248	14.36±0.09	0.80±0.08	1.40±0.08	1.840±0.07	0.187	0.35770±0.00001	~0.22 (APACHE I band)	1632.60	EW	~17.6

Notes: The B and V magnitudes (and corresponding uncertainties) are taken form the UCAC4 catalog (Zacharias et al. 2013), while the J and K magnitudes (and corresponding uncertainties) come from the 2MASS survey (Skrutskie et al. 2006). The acronyms used for the classification of the variables follow the variable star type designations of the AAVSO International Variable Star Index (VSX: Watson et al. 2014; see http://www.aavso.org/vsx/help/VariableStarTypeDesignationsInVSX.pdf).

a: Celestial coordinates are taken from the UCAC4 catalog (Zacharias et al. 2013).
b: Integrated along the line of sight.
c: Possible secondary minimum visible in the APACHE data, with the folded light curve appearing distorted.

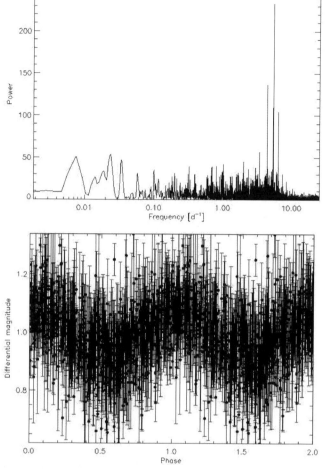

Figure 1. Star UCAC4 837-000728. (Upper plot) Lomb-Scargle periodogram of the APACHE photometric data. (Lower plot) APACHE light curve folded at the peak period P = 0.185848 day.

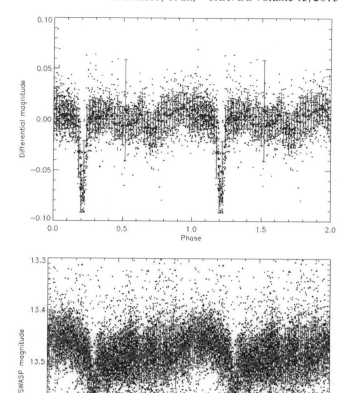

Figure 2. Eclipsing binary UCAC4 612-044588. (Upper plot) APACHE light curve (I band), folded according to the best orbital period P = 0.88965 day. (Lower plot) SuperWASP light curve, 3-sigma clipped and folded according to the same period.

The star was also observed by the SuperWASP survey (more than 6600 data collected between September 23, 2004, and May 18, 2008), but the light curve (lower panel of Figure 2) is characterized by a higher scatter than the APACHE time series, then making difficult the detection of the secondary minimum. The light curve distortion is also evident in the SuperWASP data and it is probably due to a direct physical interaction between the two components caused by their proximity.

3.3. UCAC4 667-058562

The APACHE data for this target cover the quite limited timespan March 7–May 15, 2014. They show variations in the light curve that appear to have some periodicity that can be related to the rotation of the star (Figure 3). In such a case, the flux modulation is produced by inhomogeneities in the stellar photosphere, as spots and active regions, that trace the mean rotation period of the star. By applying the Generalized Lomb-Scargle (GLS) algorithm (Zechmeister and Kürster 2009), we found a peak in the periodogram at f_{max} ~0.0941 cycle/day (corresponding to P ~10.62 days), and the data can be reliably fit with a sinusoid of semi-amplitude ~0.07 mag. Fitting the light curve of a star, which shows evidence of rotation, with a single sinusoid represents only a simplified model, because this describes the average structure of the photosphere over time and it does not take into account the life cycle of the active regions/spots and the changes occurring in their longitude distribution over the stellar disk. The nearly two-month APACHE observations suggest evidence of changes in the photospheric structures of the star over time with a time scale close to that of a single rotation cycle. This can be observed by comparing the different amplitudes of minima and maxima in the time series, reflected in the structure of the O–C residuals (upper panel of Figure 3). To quantitatively assess the significance of the GLS peak frequency, we performed a bootstrap analysis (with re-sampling) of the APACHE data, from which it is possible to guess if the observed peak is real and at which level of confidence. It consists of two steps: i) randomly shuffling the magnitudes while keeping fixed the time stamps (allowing for multiple extractions of the same data point), and ii) performing a GLS analysis on the new dataset, with the same settings used to analyze the original data. If the observed signal is real, the shuffling should destroy without producing in the periodogram a higher peak corresponding to a different frequency f. On the contrary, if it is due simply to white noise, a more significant peak should appear, at any frequency, in several fake datasets. By repeating the bootstrap N times (N = 10,000 in our case), we determined the total number of fake datasets for which the peak spectral power density, associated

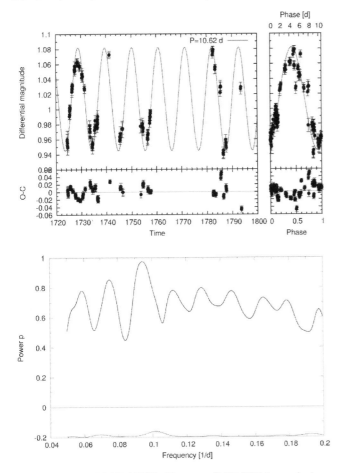

Figure 3. Star UCAC4 667-058562. (Upper panel) APACHE time series (upper left), and the folded light curve at P = 10.62 days (upper right). Correspondingly, the residuals of a sinusoidal fit for both datasets are shown in the two lower plots. Time is provided as HJD − 2455000. (Lower panel) GLS periodogram of the APACHE (blue curve). The best peak found is at f = 0.0942 cycles/day. The green curve represents the window function of the APACHE observations.

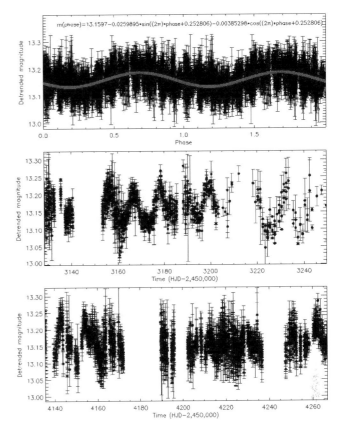

Figure 4. Light curve of the star UCAC4 667-058562 (1SWASP-J135821.53+432058.4) as observed by the SuperWASP survey. (Upper plot) Data folded according to the period P=10.62 days, superposed with the best fit function of the type $A + B \times \sin(2\pi \times phase + C) + D \times \cos(2\pi \times phase + C)$. The curve consists of 3622 data points and was obtained by binning the original dataset in bins of 0.005 day and applying a 3-sigma clipping. (Middle and lower plots) Time series spanning almost four years of observations.

to any frequency, was higher than the spectral power density associated to f_{max} in the case of the original dataset. It resulted that none of the fake datasets produced a spectral power density higher than the original, and for the P = 10.62 days signal this corresponds to a False Alarm Probability (FAP) of 10^{-4}, suggesting that it is not due to pure noise. FAP levels of 0.1% and 1%, respectively ten and one hundred times higher than that associated to f_{max}, correspond to spectral power densities p = 0.86 and p = 0.83, which are very close to the power of the secondary peak in the original periodogram at f = 0.075 cycle/day (P ~13 days). This suggests that the secondary peak is not very significant but, due to our limited dataset, it could reflect the uncertainty we have about the stellar rotation period, or it could be related to the time scale over which the evolution of the active regions occurs.

The scenario emerging from the APACHE photometry is supported by the SuperWASP observations. This survey monitored the star, named 1SWASP-J135821.53+432058.4, between May 02, 2004, and April 18, 2008, collecting 13,768 points. The SuperWASP light curve, averaged in bins of 0.005 day, is shown in Figure 4, both the folded data at P = 10.62 days and the time series. The data, characterized by typical uncertainties higher than that of APACHE and covering a much longer timespan, show changes occurring in the light curve morphology over a time scale of the order of the rotation frequency. This strengthens the hypothesis of a rapid evolution of photospheric structures as spots and active regions. The GLS analysis applied to the SuperWASP data returned a peak at f~0.090 cycle/day (P~11 days), which confirms the result of the APACHE data analysis.

3.4. UCAC4 854-011628

This object was observed by APACHE in V band between June 1 and October 21, 2014, collecting 1,405 useful points. Figure 5 shows the light curve folded at P = 0.30292 day, clearly indicating the nature of EW eclipsing binary for this target, with equal maxima and primary and secondary minima having a difference in their depths of ~0.03 mag. No data are available from SuperWASP.

3.5. UCAC4 609-091606

We observed this target between July 12 and August 27, 2012, collecting 1,030 useful data points in I band. Figure 6 shows the light curve folded according to the best, nonetheless tentative period found P = 1.738 days, by refining the result of the analysis performed with the Box Least-Squares (BLS)

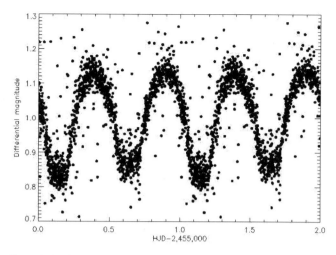

Figure 5. APACHE light curve in V band of the star UCAC4 854-011628, folded according to the period P = 0.30292 day, clearly showing the existence of primary and secondary minima and a morphology typical of an EW eclipsing binary.

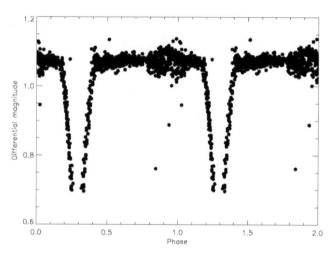

Figure 6. APACHE light curve of the star UCAC4 609-091606 folded according to the tentative period P = 1.738 days.

algorithm (Kovács *et al.* 2002). While the nature of eclipsing binary appears clear, from our data alone we cannot confirm the found periodicity, which could possibly be twice our estimate. Due to lack of data we cannot fully characterize the minimum we have observed but only provide an upper limit to its real depth, and we cannot explain the cause of the scattered portion of the light curve visible in the phase range [0.8, 1], which we could exclude to be due to instrumental noise and could actually be related to the existence of a secondary minimum. The solution corresponding to a period P = 3.476 days is reliable, but the phase coverage of our data is not enough to prefer this hypothesis over the other. Unfortunately, no SuperWASP data are available to assess the real orbital period of this eclipsing binary.

3.6. UCAC4 610-092815

APACHE observed this object in I band between July 12, 2012, and August 14, 2014, for a total of 1,422 useful points. No SuperWASP observations are available. This variable appears to be an EW eclipsing binary, for which we estimate an orbital period P = 0.42706 day. The folded APACHE light curve is shown in

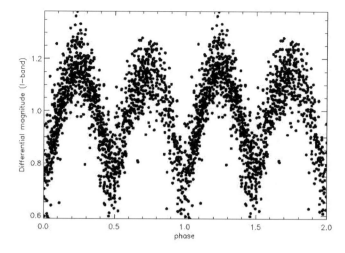

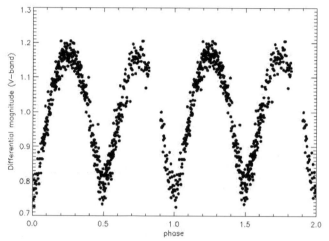

Figure 7. Star UCAC4 610-092815. (Upper plot) APACHE light curve folded according to the orbital period P = 0.42706 day. (Lower plot) Data collected with the OAVdA 81-cm telescope in V band and folded according the same ephemeris.

Figure 7. Despite the light curve appearing scattered due to the faint magnitude of the star (V = 15.741), because the observations with a 40-cm telescope were not optimized for this system, the two maxima show a different height, while the two minima appear to have the same depth (within the scattering level).

Due to a follow-up campaign which was focused on a different target from UCAC4 610-092815 but in the same field of view, we collected photometric data for this EW system in V band with the OAVdA 81-cm telescope. The observations were carried out during the nights of 13, 14, and 15 August 2014, for a total of 566 useful data. The corresponding folded light curve is shown in the lower plot of Figure 7. Due to the bigger aperture of the telescope, the data show much less scatter than those obtained in I band and confirm that the two maxima have not the same height, while the minima appear of the same depth also in V.

3.7. UCAC4 621-119831

We observed this star in V band between June 21 and September 28, 2014, for a total of 1,011 useful points. The photometric time series is shown in Figure 8, from which it appears that the variable changes its magnitude in an irregular way, at least from our dataset. We tentatively classify this object

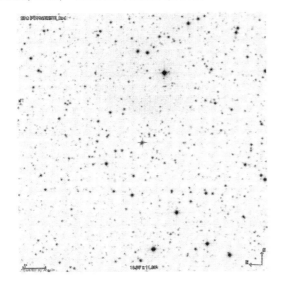

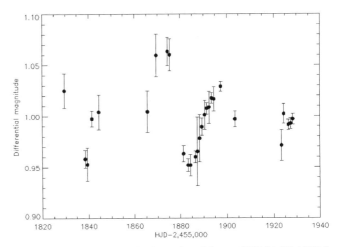

Figure 9. APACHE photometric time series of the star UCAC4 620-119316. Each point represents the average of the data collected during a single night of observation, and the error bars are the r.m.s. of the data of the corresponding night.

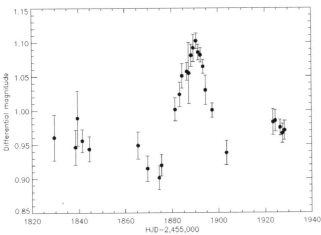

Figure 8. (Upper figure) Sky chart indicating the object UCAC4 621-119831 (red cross in the middle of the image). The image was downloaded from the Aladin sky atlas service and has linear dimensions 10.96' × 11.08'. The image scale (1 arcsec) is indicated in the lower left corner. (Lower plot) APACHE photometric time series. Each point represents the average of the data collected during a single night of observation, and the error bars are the r.m.s. of the data of the corresponding night.

as L variable. The flux of the star appears to be red-dominated (B–V = 1.757, V–Ks = 4.759, J–H = 0.909, H–Ks = 0.268), with a reddening weighing in at level of a tenth (or less) of magnitude. The color indexes are compatible with those of an early M giant (Zombeck 1990), and appear not compatible with those of a red dwarf. Without a reconnaissance spectrum available, our analysis is necessarily based on the expected intrinsic colors. Taking as a reference the intrinsic colors derived by Pecaut and Mamajeck (2013) for main sequence stars (see Table 4 therein), it can be seen for example that the combination of the V–J, V–K, and J–H color indexes is not representative of an M dwarf, also taking into account their uncertainties, which are of the order of few hundredths of magnitude (Table 1).

Looking at the finding chart (Figure 8), we note that UCAC4 621-119831 is very close to the much fainter star UCAC4 621-119835, which is separated by ~5" from the target, or ~3.5 pixels in one APACHE image. No measurements of the V magnitude are available in the VizieR archive for this visual companion, which is ~5 magnitudes fainter in J band (2MASS photometry). Our data are not corrected for blending, but we believe that, considered the large difference in luminosity between the pair of objects, the level of contamination should be very negligible in this case.

No SuperWASP data are available for this object.

3.8. UCAC4 620-119316

This star shares the same field of UCAC4 621-119831 in the APACHE scientific frames, thus it was observed in the same time span, collecting 954 useful points. The APACHE time series is shown in Figure 9. The star appears as a single star, with no risk of blending with another object, and its flux is red-dominated, with color indexes similar to those of UCAC4 621-119831 (B–V = 1.842, V–Ks = 4.723, J–H = 0.94, H–Ks = 0.246). For the same reasons discussed for the previous variable, the color indexes do not appear compatible with the intrinsic values expected for a M dwarf star, suggesting this also should be an M giant. The light curve is indicative of an L irregular variable.

3.9. UCAC4 620-119722

This star was observed by APACHE in V band between June 21 and September 28, 2014, for a total of 974 useful points. We found photometric data also in the SuperWASP archive, covering the period between June 23 and October 26, 2007, for a total of 2198 points. The Lomb-Scargle algorithm results in almost the same periodogram for both the datasets, with a best period of 0.2778 day. The folded light curves (see Figures 10 and 11) appear to have a quite sinusoidal shape, with data from APACHE much less scattered than those of SuperWASP. We note that for the SuperWASP dataset the second more relevant peak in the L-S periodogram (at 0.3850 day) is the third significant peak in the APACHE periodogram, while the third SuperWASP peak (at 0.2173 day) is the second for relevance for the APACHE dataset.

By looking at the index colors listed in Table 1, and taking into account the non-negligible, integrated reddening along the line of sight, this variable could be tentatively classified as an

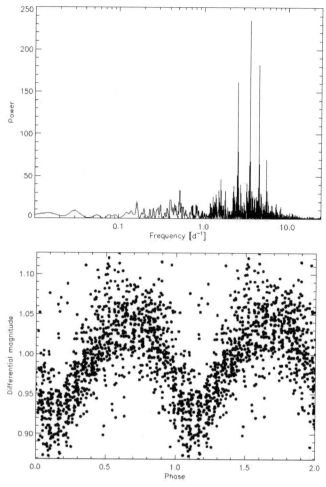

Figure 10. Star UCAC4 620-119722. (Upper plot) Lomb-Scargle periodogram for the APACHE data. (Lower plot) APACHE data folded according to the period 0.2778 day.

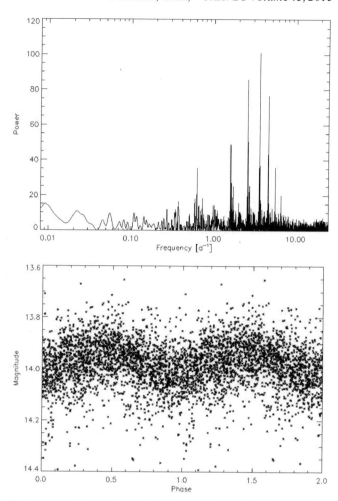

Figure 11. (Upper plot) Lomb-Scargle periodogram for the SuperWASP data of the star UCAC4 620-119722. The best period coincides with that found in the APACHE data. (Lower plot) SuperWASP light curve folded according to the best period 0.2778 day.

early-mid F-type main sequence star (Zombeck 1990; Pecaut and Mamajeck 2013). The spectral type, the periodicity and the amplitude of the light curve variations are compatible with those of a δ Scuti, which we then propose as a reliable classification.

3.10. UCAC4 673-106048

We observed this star between August 28, 2013, and September 28, 2014, for a total of 797 useful measurements in I band. The APACHE photometry is shown in Figure 12, revealing a clear variation in the star luminosity with no periodic modulation. For the same reasons previously discussed, this star is characterized by color indexes that appear to be not representative of a dwarf star, but rather they are more typical of a M giant (B–V = 1.92, V–K = 7.281, J–H = 0.996, H–Ks = 0.423). In particular, while the V–K index is compatible with that of a M6V star, the observed V–J index appears ~0.5 mag. lower than expected value (and this difference is expected to be even higher, if we consider the reddening correction that lowers the V–J index), and J–H is ~0.35 mag. higher than the tabulated value. We therefore classify it as an LB-type (slow irregular of late spectral type) pulsating variable.

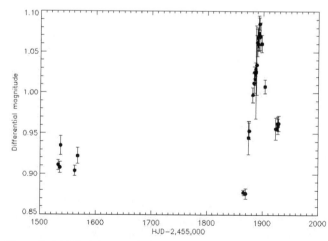

Figure 12. APACHE photometric time series of the star UCAC4 673-106048. Each point represents the average of the data collected during a single night of observation, and the error bars are the r.m.s. of the data of the corresponding night.

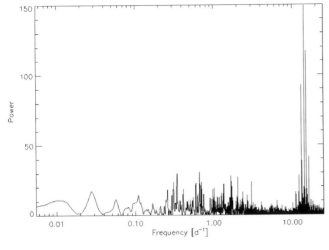

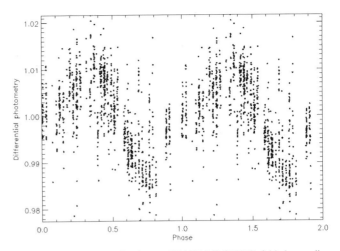

Figure 15. APACHE data for the star UCAC4 848-018678, folded according to the period P = 41 days.

3.11. UCAC4 858-013784

APACHE observed this star between April 3 and October 2, 2013, with 1,574 useful data in I band. Figure 13 shows the Lomb-Scargle periodogram of our measurements characterized by a peak at 0.07077 day, and the light curve folded at this period, characterized by amplitude of ~0.02 mag. The color indexes (see Table 1, and also J–H = 0.177, H–Ks = 0.059), taking into account the magnitude of the integrated reddening term E(B–V), appear to be compatible with those of a late F-type main sequence star (for example, Pecaut and Mamajeck 2013). Together with the light curve characteristics, this supports the classification of the variable as a δ Scuti star. No SuperWASP data are available.

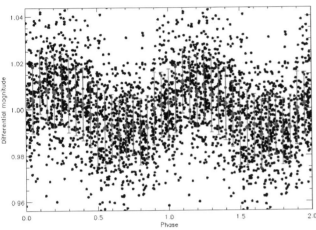

Figure 13. Star UCAC4 858-013784. (Upper plot) Lomb-Scargle periodogram for the APACHE data. (Lower plot) APACHE data (complete dataset in black and average values within bins of 0.05 in red) folded according to the period 0.07077 day.

3.12. UCAC4 849-017521

This object was observed in I band from March 16, 2013, to June 2, 2014, for a total of 2,833 measurements. We applied the BLS algorithm to the light curve obtained by normalizing the measurements of each single night to their median value. This variable was found to be an eclipsing binary of EW type, with an orbital period P = 0.5988 ± 0.0001 day. We show in Figure 14 the light curve binned at 0.005 day and folded according to the period found, which clearly reveals the presence of a primary and a secondary minimum with different depths, while the two maxima appear equal within the scatter of our data. No SuperWASP data are available for this variable.

3.13. UCAC4 848-018678

This star is in the same field of view of the previous variable, and we collected 2,821 useful measurements of it during the same time span. Figure 15 summarizes our results. By applying the GLS algorithm we found a peak periodicity at ~41 days. We folded the APACHE data to this period (Figure 15), from which it appears that the light curve follows a sinusoidal-like modulation with ~0.02 peak-to-valley amplitude. This can be interpreted as a photometric variation related to the stellar rotation and due to an unevenly spotted photosphere. Due to our sparse measurements, nonetheless collected during a time span of nearly 15 months, we cannot be very confident that the true period is actually the one that we derived, which

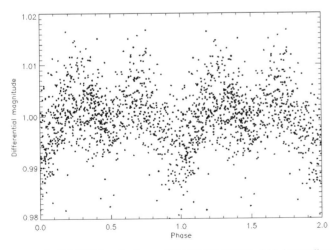

Figure 14. APACHE data for the star UCAC4 849-017521, folded according to the 0.5988-day period.

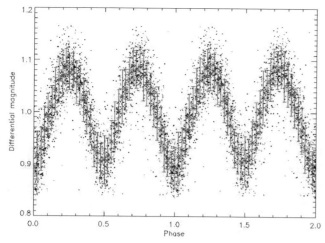

Figure 16. APACHE data for the star UCAC4 849-017658, folded according to the period 0.35770 day, with superposed the binned light curve in red (one bin corresponding to an interval of 0.025 in phase).

is quite long and would require the monitoring of several consecutive rotations to be better constrained. In absence of any other information about this star, such as mass, radius, and age, we cannot use gyrochronology relations to support our interpretation. From the color indexes and color excess (see Table 1, and also J–H = 0.701, H–Ks = 0.223) we suggest that the star could be tentatively classified as an early-type M dwarf (Zombeck 1990; Pecaut and Mamajeck 2013). Photospheres of M dwarfs are generally expected to be spot-covered, thus rotational modulations in the light curve are frequently detected, as we found for some of the APACHE targets. The UCAC4 catalog (Zacharias et al. 2006) gives pmRA = –0.4 ± 2.3 mas/yr and pmDE = 1.4 ± 2.6 mas/yr for the proper motion of this star, which is indeed not indicative of a nearby star. Therefore, another reliable possibility is that the star is actually a red giant, for which a small proper motion is highly probable. The color indexes also support this scenario, being compatible with a late KIII/early MIII star (Zombeck 1990). If this is the case, this star could be assigned to the group of the semiregular (SR) variables showing small-amplitude variations in its light curve.

3.14. UCAC4 849-017658

Found in the same field of the previous three variables, we collected for this object 2,824 useful measurements in the I band. By using the Lomb-Scargle algorithm we found half of the real period to have the highest peak in the periodogram. By doubling this value and folding the light curve accordingly (P = 0.35770 day), we recognized that its shape is clearly that of a short period eclipsing binary of EW type (Figure 16). While the two maxima appear of the same height, taking into account the scatter in our data, the minima have slightly different depths (Δm ~0.015 mag.), as shown by the binned curve.

4. Acknowledgements

This research has made use of several resources: the VizieR catalogue access tool and the SIMBAD database, operated at CDS, Strasbourg, France. We also used data from the first public release of the WASP data (Butters *et al.* 2010) as provided by the WASP consortium and services at the NASA Exoplanet Archive, which is operated by the California Institute of Technology, under contract with the National Aeronautics and Space Administration under the Exoplanet Exploration Program. Data were also used from the NASA/IPAC Infrared Science Archive, which is operated by the Jet Propulsion Laboratory, California Institute of Technology, under contract with the National Aeronautics and Space Administration.

MD acknowledges partial support from INAF-OATo through the grant "Progetto GAPS: caratterizzazione spettroscopica e fotometrica dei target (attività cromosferica, rotazione) e studio delle sinergie tra GAPS e APACHE" (#35/2014), and from ASI under contract to INAF I/058/10/0 (Gaia Mission—The Italian Participation to DPAC). JMC and AB are supported by a grant of the European Union-European Social Fund, the Autonomous Region of the Aosta Valley and the Italian Ministry of Labour and Social Policy. We thank ASI also through contract I/037/08/0, and Fondazione CRT for their support of the APACHE Project.

The Astronomical Observatory of the Autonomous Region of the Aosta Valley is supported by the Regional Government of the Aosta Valley, the Town Municipality of Nus, and the Mont Emilius Community.

References

Butters, O. W. *et al.* 2010, *Astron. Astrophys.*, **520**, L10.
Christille, J.-M., *et al.* 2013, in *Hot Planets and Cool Stars*, ed. R. Saglia, EPJ Web of Conferences 47, id.17001 (http://publications.edpsciences.org/).
Damasso, M., *et al.* 2014, *J. Amer. Assoc. Var. Star Obs.*, **42**, 99.
Kovács, G., Zucker, S., and Mazeh, T. 2002, *Astron. Astrophys.*, **391**, 369.
Pecaut, M. J., and Mamajek, E. E. 2013, *Astrophys. J., Suppl. Ser.*, **208**, 9.
Scargle, J. D. 1982, *Astrophys. J.*, **263**, 835.
Schlafly, E.F., and Finkbeiner, D.P. 2011, *Astrophys. J.*, **737**, 103.
Skrutskie, M.F., *et al.* 2006, *Astron. J.*, **131**, 1163.
Sozzetti, A., *et al.* 2013, in *Hot Planets and Cool Stars*, ed. R. Saglia, EPJ Web of Conferences 47, id.03006 (http://publications.edpsciences.org/).
Watson, C., Henden, A. A., and Price, C. A. 2014, AAVSO International Variable Star Index VSX (Watson+, 2006–2014; http://www.aavso.org/vsx).
Zacharias, N., Finch, C. T., Girard, T. M., Henden, A., Bartlett, J. L., Monet, D. G., and Zacharias, M. I. 2013, *Astron. J.*, **145**, 44.
Zechmeister, M., and Kürster, M. 2009, *Astron. Astrophys.*, **496**, 577.
Zombeck, M. V. 1990, *Handbook of Space Astronomy and Astrophysics*, 2nd ed., 1990, Cambridge Univ. P., Cambridge.

UXOR Hunting among Algol Variables

Michael Poxon
9 Rosebery Road, Great Plumstead, Norfolk NR13 5EA, United Kingdom; mike@starman.co.uk

Received February 18, 2015; revised March 31, 2015; accepted May 20, 2015

Abstract The class of variable typified by UX Orionis (UXORs or UXors) are young stars characterised by aperiodic or semiperiodic fades from maximum. This has led to several of the class being formerly catalogued as Algol-type eclipsing binaries (EAs), which can show superficially similar light variations. With this in view, I propose a campaign to search for more UX Ori type stars.

1. History

UX Orionis is the prototype of a subset of Young Stellar Objects (YSOs) which spend most of their time at or near maximum brightness, but which undergo periods of fading which can vary in any of: 1. amplitude; 2. duration; or 3. regularity.

No UXORs are known that exhibit amplitude, duration, or regularity as predictably as do typical Algol-type eclipsing stars, even though in the past several objects presently classified as UXORs were erroneously listed as EA-type variables, due largely no doubt to incomplete or inadequate observations. An example is BO Cephei, of which Wenzel (1991) said:

> The reader must be aware of the fact that obviously the minima are of very different depth and that moreover in numerous cases they cannot be realized at all at the predicted dates.... A thorough spectroscopic investigation of this important object is clearly overdue. In this connection we remind (sic) of two more kinds of "eclipsing" stars with vanishing minima, represented by Kohoutek's central star V 651 Mon of a planetary nebula.... Maybe these cases are more numerous than hitherto supposed. The large number of eclipsing variables, for which no period could be found, could be explained in this way....

When BO Cep was discovered (Morgenroth 1934) it was classified as an Algol (with a rather scary range of 14–<15.5) but by the time of Wenzel's article its true nature was of course known.

It is also possible for UXORs to be allocated other types; for instance, GT Ori in Table 1 was originally thought to be an SRd variable, presumably on its F0 spectrum and dearth of observational evidence. UXORs may also be confused with the R Coronae Borealis stars, which show similar aperiodic fades, as well as another type of young object typified by FU Orionis, previously lumped under the "slow novae" category though totally unrelated to novae, an example being the star CB 34V (Tackett and Herbst 2003). In this campaign, however, I am concentrating purely on the EA stars.

2. Astrophysical background

UXORs are drawn almost exclusively from early spectral type stars with masses $>2_\odot$ as distinct from, say, the T Tauri objects, which tend to be of types G to M and of masses comparable to, or slightly less than, the Sun. One peculiarity of UXORs is that when a fade has taken place the spectral signature moves bluewards. This phenomenon tends to support the model of an UXOR as a star surrounded by an accretion disc possibly consisting of accreting planetesimals, with the disc obscuring the actual star, scattering its light, with the shorter-wavelength blue light predominating—essentially the same phenomenon that produces blue skies here on Earth. Many physical factors come into play at this stage, chief of which is probably the inclination of the system with respect to the observer. The lesser amount of light contributed by the star itself means that the disc provides more (scattered) light, which contributes to the blueing effect. Inclinations approaching 90° should therefore show no excess of scattered light over starlight, since both star and disc continue to contribute to the total light output, but also for this reason there should be no fades seen either, unless the star itself is independently variable, which is a possibility. In addition, since the disk is obscuring the star at such an inclination, the light source itself will be very faint, and only the scattered light will be seen. Clearly, however, there will be a whole "spectrum" of variational behavior from this cause since we can expect inclinations to be randomly distributed. The wide variety of amplitudes can be seen in Table 1, which shows those stars currently considered to be UXORs.

Most star formation occurs in discrete areas of the sky where progenitor molecular clouds occur. Stars formed in these regions have become known as T-associations or OB-associations, depending on whether the main variables found there are the low-mass T Tauri stars or the more massive early-type variables, respectively; it is the latter type that concerns us here, although in practice both types may be found in the same region. These associations are named from the constellations where they are found, thus Ori T2 is that starforming region centred on the Orion nebula and includes our old friend T Ori. Another is located around λ Ori and includes several AAVSO program stars such as CO Ori. As can be seen in Table 1, both T Ori and CO Ori are UXORs, but occur in areas where T Tauri stars also occur, GW Ori, for instance, in the case of CO Ori, and a host of others in the case of the M42 variables.

A third type of stellar association is that associated with reflection nebulae, the R-association. These have B-type stars as their earliest spectral type members, and there are notable examples of such regions in Monoceros, Canis Major, and Vela.

Table 1. Stars presently classified as UX Orionis type (VSX; Watson et al. 2015).

Star	R.A. (2000) h m s	Dec. (2000) ° ′ ″	Range
ASAS J072505-2545.8	07 25 04.95	–25 45 49.6	12.6–15.4 V
VX Cas	00 31 30.69	+61 58 51.0	10.5–13.3 V
MQ Cas	00 09 37.56	+58 13 10.7	10.8–13.9 V
NSV 5178	11 22 31.67	–53 22 11.5	6.75–6.79 V
SV Cep	22 21 33.21	+73 40 27.1	10.35–12.15 V
BG Cep	22 00 30.64	+68 28 22.8	13.2–14.3 p
BH Cep	22 01 42.86	+69 44 36.5	10.79–12.7 V
BO Cep	22 16 54.06	+70 03 45.0	11.5–12.4 V
BS Cep	22 29 05.43	+65 14 41.9	13.9–16.0 p
GM Cep	21 38 17.32	+57 31 22.0	12.9–15.1 V
IL Cep	22 53 15.61	+62 08 45.0	9.24–9.61 V
LO Cep	21 19 43.01	+61 42 26.6	13.4–15.0 V
V373 Cep	21 43 06.81	+66 06 54.1	11.82–13.3 V
ASAS J152008-6148.4	15 20 07.38	–61 48 27.7	12.1–13.9 V
V517 Cyg	20 47 23.59	+43 44 39.8	12.1–15.6 V
V1686 Cyg	20 20 29.35	+41 21 28.4	12.5–17.2 V
V1977 Cyg	20 47 37.47	+43 47 25.0	10.8–11.8 V
NSV 20441	15 56 41.89	–42 19 23.3	8.27–8.60 V
VY Mon	06 31 06.93	+10 26 05.0	12.8 V–17.4 p
IRAS 06068-0643	06 09 13.70	–06 43 55.6	14.5–20.1 CV
KR Mus	11 33 25.44	–70 11 41.2	6.67–6.92 V
ASAS J172056-2603.5	17 20 56.13	–26 03 30.7	12.7–13.9 V
T Ori	05 35 50.45	–05 28 34.9	9.5–12.6 V
UX Ori	05 04 29.99	–03 47 14.3	9.48–12.5 V
BF Ori	05 37 13.26	–06 35 00.6	9.69–13.47 V
CO Ori	05 27 38.34	+11 25 38.9	10.0–12.8 V
GT Ori	05 43 29.25	+00 04 58.9	10.6–13.4 V
HK Ori	05 31 28.04	+12 09 10.3	11.2–12.3 V
V350 Ori	05 40 11.77	–09 42 11.1	10.57–13.5 V
V586 Ori	05 36 59.25	–06 09 16.3	9.48–11.41 V
V1012 Ori	05 11 36.55	–02 22 48.5	11.8–<14.2 V
ASAS J055007+0305.6	05 50 07.14	+03 05 32.5	11.1–13.1 V
NSV 16694	05 50 53.72	+03 07 29.4	11.8–13.4 V
RZ Psc	01 09 42.05	+27 57 01.9	11.25–14.2 V
NX Pup	07 19 28.26	–44 35 11.3	9.0–11.2 V
V718 Sco	16 13 11.59	–22 29 06.6	8.75–10.30 V
V856 Sco	16 08 34.29	–39 06 18.4	6.77–8.0 V
V1026 Sco	15 56 40.02	–22 01 40.0	8.57–9.5 V
NSV 8338	17 13 57.44	–33 07 46.0	12.4–<16 V
XX Sct	18 39 36.99	–06 43 05.7	13.0–16.6 p
GSC 05107-00266	18 27 26.08	–04 34 47.5	10.96–13.2: V
VV Ser	18 28 47.87	+00 08 39.9	11.5–<13.6 V
RR Tau	05 39 30.51	+26 22 27.0	10.2–14.3 V
CQ Tau	05 35 58.47	+24 44 54.1	8.7–11.9 V
HQ Tau	04 35 47.33	+22 50 21.7	12.1–14.5 V
EM Vel	08 35 40.30	–40 40 07.2	11.7–13.2 V
FX Vel	08 32 35.77	–37 59 01.5	9.4–11.5 V
WW Vul	19 25 58.75	+21 12 31.3	10.25–12.94 V
PX Vul	19 26 40.25	+23 53 50.8	11.4–12.8 V

UXORs tend to be of B to F spectra and so these areas could be considered as prospective hunting grounds also.

3. Observational possibilities

Table 1 reveals that many of the presently-known UXORs are reasonably bright objects, indeed, AB Aur is visible entirely with binoculars (though it does not appear in Table 1 because it is not simply an UXOR type). However, simple "eyepiece" observation of stars thought to be Algol types is a highly unprofitable exercise for several reasons:

1. there are far too many EA stars to monitor (VSX lists over 20,000);

2. breaks in observation due to weather or other circumstances;

3. some stars may have unsuitable parameters (too faint, too small amplitude).

With this in mind, we need to perform some serious pruning. Most starforming regions lie on or near the galactic equator since that is where most of the progenitor material is found, and so the likelihood of finding an UXOR in these areas increases. It is true that in recent years we have found stellar associations away from the galactic plane—for example in the far-southern constellation of Chamaeleon; it remains true that to unearth an UXOR means that we should stick to the most active areas of the sky. Therefore I have drawn up a list of stars currently assumed to be Algol-type systems that lie within 5°

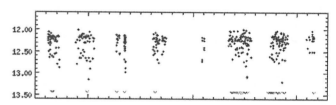

Figure 1. A typical light curve of AL Tau, an Algol star with a period of about a day but which lies in a starforming region close to such UXORs as RR Tau and CQ Tau. The x-axes in this and in Figure 2 are unlabelled since what we are interested in here is not the period, but merely the presence of a regularity; and indeed there appears to be a strong regularity here, so we can assume that AL Tau is not an UXOR. Light curve data are from ASAS3 (Pojmański et al. 2013).

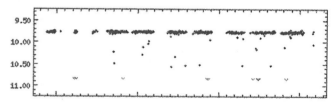

Figure 2. Shows a typical light curve for FP Car, again located near a starforming region. While the fades do not appear to be as regular as those for AL Tau (there are also fewer data points) note that the "fainter than" symbols do seem to occur at reasonably regular intervals; one can mentally insert a hypothetical "fainter than" where it appears to be absent. But for these reasons maybe we should not be in such a hurry to drop this star from our list as in the previous example. Light curve data are from ASAS3 (Pojmański et al. 2013).

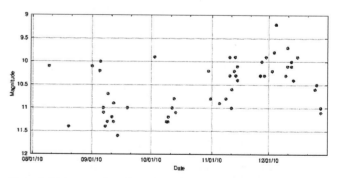

Figure 3. Light curve for a known, and highly-active, UXOR (CQ Tau) for comparison. This time we are using AAVSO data from the AAVSO Light Curve Generator, and it can immediately be seen that while there may be a vague periodicity to some minima, the fades when they occur are not always of the same depth or duration. This should be enough to tell us that we are not dealing with an Algol-type star.

of the galactic equator. This provides us with a list of 357 stars. Still a fair few, but a step down from 20,000 objects!

We can narrow this list down even more if we consider the light curves of selected stars. We need to look for the presence of strict regularity, which will strongly suggest that we are dealing with a genuine Algol variable, and so can eliminate it from our search list. A good resource for this is ASAS (Pojmański *et al.* 2013; http://www.astrouw.edu.pl/asas/?page=aasc&catsrc=asas3) although currently curves are only available for stars South of $+28°$.

Figures 1 through 3 are a selection of curves illustrating what can be seen.

4. Summary: future of the campaign

As of the present (March 2015) several observers have expressed an interest in following the UXOR hunt and have been provided with a list of the stars discussed above. There are now search facilities on the YSO section website (http://www.starman.co.uk/ysosection) via a queryable MySQL database where observers can filter targets to suit their needs.

In order for the central list to be trimmed even more it would be useful if any details involving the existing candidates (definitely proven Algol type, and so on) could be communicated to me so that I can remove them from the database.

From what was discussed above, it is also probably a truism to say that EA stars with accurately-determined periods which can be shown to have a high degree of regularity should indeed be considered to be bona fide Algols and therefore eliminated from the original list—although there are a very small number of stars such as KH15D that exhibit a high degree of regularity while being a YSO rather than an Algol star (see for example Kearns and Herbst 1998). However, to compensate for this, we may want to consider not only those EA stars within $5°$ of the galactic equator but also those in R-associations, for the reasons pointed out earlier.

The list currently comprises all types of variation within the EA parameters—a wide range of periods, amplitudes, and brightness levels—so it will be coverable by visual and CCD observers, ensuring that a comparatively large number of observers can participate. In the near future I would also like to add some additional stars in the region of certain R-associations, in which regard I would like to thank Prof. Bill Herbst for his helpful suggestions.

References

Kearns, K. E., and Herbst, W. 1998, *Astron. J.*, **116**, 261.
Morgenroth, O. 1934, *Astron. Nachr.*, **252**, 389.
Pojmański, G., Szczygiel, D., and Pilecki, B. 2013, The All-Sky Automated Survey Catalogues (ASAS3; http://www.astrouw.edu.pl/asas/?page=aasc&catsrc=asas3).
Tackett, S., and Herbst, W. E. 2003, *Astron. J.*, **126**, 348.
Wenzel, W. 1991, *Inf. Bull. Var. Stars*, No. 3647, 1.
Watson, C., Henden, A. A., and Price, C. A. 2014, AAVSO International Variable Star Index VSX (Watson+, 2006–2015; http://www.aavso.org/vsx).

Sudden Period Change and Dimming of the Eclipsing Binary V752 Centauri

Anthony Mallama
14012 Lancaster Lane, Bowie, MD 20715; anthony.mallama@gmail.com

Hristo Pavlov
9 Chad Place, St. Clair, NSW 2759, Australia; hristo_dpavlov@yahoo.com

Received February 5, 2015; revised March 1, 2015; accepted May 12, 2015

Abstract Video photometry was used to determine a time of minimum light of V752 Centauri at heliocentric JD 2457051.1458 ± 0.0002. The observed time was late by nearly two hours relative to the ephemeris in the *General Catalouge of Variable Stars* (GCVS). Analysis of this and other timings indicates that the orbital period of V752 Cen increased abruptly by 7.49×10^{-6} days in 2004 after remaining constant for at least the previous 34 years. Simultaneously, the star became fainter by 0.07 magnitude in the V band during primary eclipse, as indicated by an analysis of ASAS data. Dimming of 0.02 to 0.03 magnitude occurred at other phases in the light curve, too. By 2008 the star had returned to its normal brightness. The dimming and the period change may both have been the consequence of material escaping from one of the stars. The deepening of the primary eclipse is also consistent with a color change.

1. Introduction

V752 Centauri is a W UMa-type eclipsing binary star. The V magnitude range is 9.1 to 9.66 and the spectral type is F8V according to the AAVSO International Variable Star Index (VSX; Watson *et al.* 2014). The ephemeris for primary minima according to the *General Catalogue of Variable Stars* (GCVS; Kholopov *et al.* 1985) is given in Equation 1.

$$\text{JD hel} = 2444243.6916 + 0.37022484 \times N \qquad (1)$$

where N is the number of cycles. Times of minimum of V752 Cen determined prior to the year 2000 are in good agreement with this ephemeris. A secondary eclipse of V752 Cen was recorded with video equipment in 2015 by one of the authors (HP). The method of recording and analyzing video photometry will be described in a separate article. The time of minimum was determined to be at heliocentric JD 2457051.1458 ± 0.0002. This value was late by about 1.9 hours relative to the GCVS ephemeris, which alerted us that a period change had probably occurred.

2. Period study

There are many times of minimum of V752 Cen listed in the AAVSO O–C database (Nelson 2014) from 1970 through 1995 and three additional times between 2009 and 2012. In order to fill the 14-year gap from 1995 until 2009 we analyzed ASAS photometry (Pojmański 1997). The new times of minimum and O–C residuals derived from ASAS data are listed in Table 1. We found that the O–C was changing rapidly during this interval and so we separated the data into 3 groups spanning 1,000 days each.

The orbital period of V752 Cen increased abruptly by 7.49×10^{-6} days around cycle 24,000 of the GCVS ephemeris (year 2004), as shown in Figure 1. This is a large change for a star whose period had been constant since at least 1970. The updated ephemeris given in Equation 2 is based on the most recent time of primary minimum and the best fitting period after 2004.

Table 1. V752 Cen times of minimum and residuals to the GCVS ephemeris derived from ASAS photometry.

Interval*	Phase	JD (hel.)	Cycle	O–C
1	Primary	2452546.3492	22428	–0.0047
1	Secondary	2452546.5368	22428.5	–0.0022
2	Primary	2453524.1272	25070	+0.0095
2	Secondary	2453524.3073	25070.5	+0.0045
3	Primary	2454531.1528	27790	+0.0236
3	Secondary	2454531.3398	27790.5	+0.0255

*Intervals: No. 1, JD 2452000.0–2453000.0; No. 2, JD 2453000.0–2454000.0; No. 3, JD 2454000.0–2455000.0.

The O–C residuals to Equation 2 for the times of minimum recorded after the period changes are shown in Figure 2.

$$\text{JD hel} = 2456108.344 + 0.37023233 \times N \qquad (2)$$

3. Light curve anomaly

Light curves of V752 Cen for the three intervals of ASAS data are plotted in Figure 3. The rightward shift of the times of minimum with interval number is very evident. Also notice the overall brightness decrease associated with interval 2, which corresponds to the epoch of the period change. Table 2 lists the V magnitude by phase and by ASAS interval. The average V magnitude at primary eclipse during intervals 1 and 3 was 9.66, which is in perfect agreement with the VSX value. However, during interval 2 (which coincides with the epoch of the period change) the primary eclipse was 0.07 magnitude fainter than the VSX value.

The dimming during primary eclipse suggests that the secondary star was fainter at the same time that the period change occurred. Brightness decreases are also evident at other phases during interval 2 but they range from just 0.02 to 0.03 magnitude. The dimming and the period change may both have been the consequence of material escaping from one of the stars, as often happens in close binary systems. D. Terrell

Table 2. V752 Cen maximum and minimum brightness during the three ASAS intervals.

Interval*	Phase	Mag.	Phase	Mag.	Phase	Mag.	Phase	Mag.
1	Pri	9.67	~0.25	9.15	Sec	9.67	~0.75	9.14
2	—	9.73	—	9.16	—	9.69	—	9.19
3	—	9.65	—	9.13	—	9.66	—	9.18
2−(1+3)/2	—	+0.070	—	+0.020	—	+0.025	—	+0.030

*Intervals: No. 1, JD 2452000.0–2453000.0; No. 2, JD 2453000.0–2454000.0; No. 3, JD 2454000.0–2455000.0.

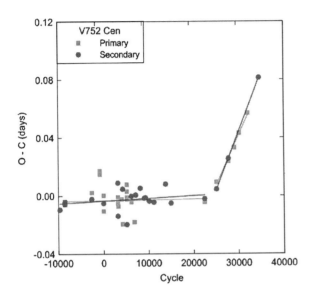

Figure 1. V752 Cen times of minimum according to the GCVS ephemeris in Equation 1 began running late after cycle 24,000, which was in year 2004.

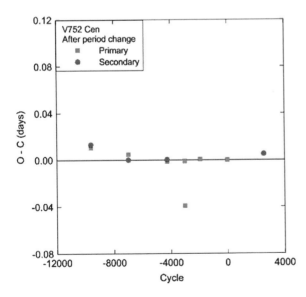

Figure 2. O–C values for the times of minimum of V752 Cen after the period change based on the updated ephemeris in Equation 2. The time of secondary minimum recorded by video and reported in this study is plotted at the far right.

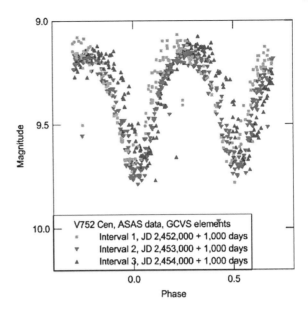

Figure 3. V-band light curve of V752 Cen compiled from three time intervals of ASAS data. The horizontal shifting with interval is due to the period increase. Notice, too, that the primary minimum was significantly fainter during interval 2.

(private communication) has noted that the deepening of the primary eclipse is also consistent with a color change.

4. Summary and conclusion

V752 Cen underwent a large orbital period increase in 2004 after a span of at least 34 years when it had not changed perceptibly. At that same time the brightness during primary eclipse dimmed by 0.07 magnitude as compared with data taken a few years before and after. Recording more times of minimum as well as full light curve photometry for this star is highly desirable. Furthermore, analysis of archival photometry for other close binaries might reveal additional light curve anomalies that have occurred simultaneously with period changes.

5. Acknowledgements

This research used information from the International Variable Star Index (VSX) database, operated at AAVSO, Cambridge, Massachusetts, USA. The authors wish to thank Dirk Terrell for reviewing an earlier version of the manuscript.

References

Kholopov, P. N., et al. 1985, *General Catalogue of Variable Stars*, 4th Ed., Moscow.
Nelson, R. 2014, AAVSO O–C database (http://www.aavso.org/bob-nelsons-o-c-files).
Pojmański, G. 1997, *Acta Astron.*, **47**, 467.
Watson, C., Henden, A. A., and Price, C. A. 2014, AAVSO International Variable Star Index VSX (Watson+, 2006–2014; http://www.aavso.org/vsx).

The δ Scuti Pulsation Periods in KIC 5197256

Garrison Turner
Big Sandy Community and Technical College, Prestonsburg, KY 41653; gturner0040@kctcs.edu

John Holaday
Purdue University Dept. of Material Engineering, Lafayette, IN 47907; jholaday@purdue.edu

Received October 21, 2014; revised January 29, 2015; accepted May 19, 2015

Abstract In this paper we present the pulsational spectrum for KIC 5197256. This object is an eclipsing binary system with a period of 6.96 days. We demonstrate that the light curve shows presence of δ Scuti pulsations with a dominant period of 0.1015 day. The object should therefore be included in the ever-growing class of eclipsing binary systems with at least one pulsating component.

1. Introduction

While the primary mission of the Kepler Space Telescope was the detection of exoplanets, several other branches of astronomy have benefited from the unprecedented photometry obtained by the instrument; perhaps none more so than that of asteroseismology. δ Scuti stars are an important component of this field, particularly if their pulsation modes can be identified. The light curves obtained from the instrument allow for detailed analysis of pulsation frequencies in a variety of regimes and thus, especially for targets which have been observed in both collection modes (LC for long cadence with integration times of ~30 minutes and SC for short cadence with integration times of ~1 minute; see Borucki (2008) for an overview of the mission and specifications), the complete pulsational behavior of certain intrinsic variables can be determined.

KIC 5197256 is a little-known system in Cygnus. It is also has the designation TYC 3139-1882-1 with coordinates R.A. $19^h 38^m 33.807^s$ and Dec. $+40° 19' 25.999"$ (J2000, simbad.u-strasbg-fr) and a visual magnitude of 11.5 (Høg et al. 2000). It has been observed in 15 quarters over the course of the Kepler Space Telescope mission with observations in both the LC and SC modes. We present the δ Scuti periods in the SC data. According to the *Kepler Eclipsing Binary Catalog* (Kirk et al. 2015) it is an eclipsing binary with a period of 6.963389 days.

δ Scuti stars reside in the HR diagram where the instability strip crosses the main sequence. Thus δ Scuti stars tend to be either main sequence or only slightly evolved stars still in the hydrogen burning phase in their life-cycle. They have typical periods between 0.02 and 0.3 day, while their amplitudes are usually on the order of millimags (although HADS, high-amplitude δ Scuti's, can have amplitudes much higher). See Breger (2000) for an overview of the characteristics of δ Scuti stars. They can pulsate in radial or non-radial modes, and are important for asteroseismology as their short periods and detectable amplitudes allow for precise period determinations with less observing time required than for variables with longer periods. On the other hand, double-lined spectroscopic eclipsing binaries allow for the determination of various orbital and component characteristics such as velocities, the semi-major axis, and the sizes, masses, and temperatures of the individual components. To date, no spectroscopy has been performed on this object. These observations are planned for a future work and should yield a more detailed picture of the orbit.

2. Observations

The Kepler Space Telescope collected one set of SC data during Quarter 4 with a start time of 2010-02-18 19:13:13 (UT) and an end time of 2010-03-19 17:07:44 (UT) (or BJD 2455246.2981 and 2455275.2116, respectively). The data were obtained from the MAST archive (AURA 2015), in which publicly available data may be retrieved. A total of 42,300 data points were used in this analysis. The light curve generated from these data are shown in Figure 1. The light curve shows definite modulation, indicative of some variability or scatter. Figure 2 shows a portion of Figure 1 on an expanded time scale to show that, while there is some scatter in the data, the main source of the modulation is pulsation modes.

3. Analysis and results

After being obtained, the data were prepared for analysis through the following steps. First, the data set included both the SAP and PDCSAP fluxes. Only the PDCSAP fluxes were used. Second, all data points included in the set taken during periods of calibration or when the telescope was offline were neglected, yielding a total of 42,325 usable data points. Third, the data were normalized by taking the average flux of all the data points and then dividing the data set through by the average. Fourth, the data were divided up into five subsets with about 8,000 data points for each subset. These subsets were then fit with a series of second-order polynomials and the fluxes divided through by the fit to remove the binarity from the light curve. The results of the fitting are shown in Figure 3 (whole light curve) and Figure 4 (the same time-span as Figure 2).

The data, once the fitting was complete, were then period-searched using the PERANSO (Vanmunster 2007) software package between 0.01 and 0.3 day to encompass the δ Scuti regime using the Lomb-Scargle method (Lomb 1976; Scargle 1982). The resulting spectrum is shown in Figure 5. The dominant period (or the period with the highest theta, the Lomb-Scargle statistic) was determined to be 0.101549 ± 0.000024 d. Figure 6 shows the polynomial-fit data phased onto this period.

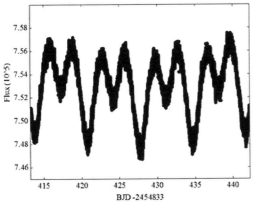

Figure 1. SC light curve from Q4 observations for KIC 5197256.

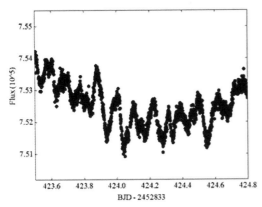

Figure 2. A portion of Figure 1 with expanded time scale to show the pulsations. Note the pulsations are visible during both eclipses.

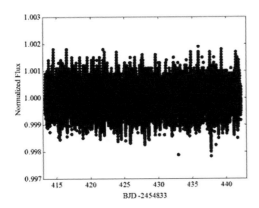

Figure 3. The entire SC light curve of KIC 5197256 after fitting with second order polynomials to eliminate orbital effects in the light curve.

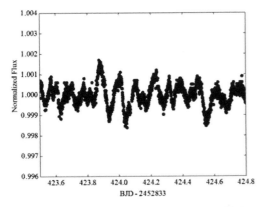

Figure 4. Shown are the data from Figure 2 after the polynomial fitting procedure was complete.

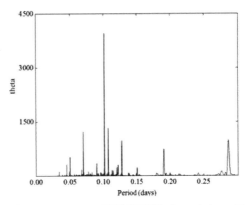

Figure 5. The power spectrum of KIC 5197256 after period searching the data between 0.01 and 0.03 day.

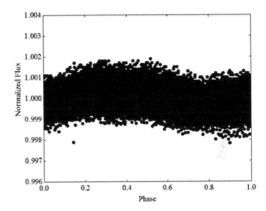

Figure 6. The normalized and fitted data of KIC 5197256 phased onto the dominant period of 0.101549 day.

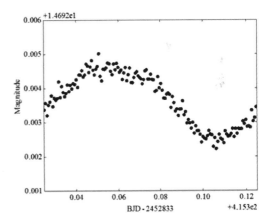

Figure 7. A small portion of data from Figure 1 set to a magnitude scale to estimate the amplitude of the dominant pulsation mode of KIC 5197256.

The ten most dominant periods found as a result of the period search are reported in Table 1 along with their associated uncertainties. It should be noted that, in an effort to ensure no true signals were lost or spurious signals gained during the polynomial fitting, the raw PDCSAP fluxes were also period-searched in the same range as the polynomial-fit data. The result is identical to Table 1.

As information about the amplitude is lost through both the normalization and the fitting, the original light curve was converted to a magnitude scale by taking the log (base 10) of the flux and then multiplying by 2.5 (the conventional minus

Table 1. Pulsational periods, uncertainties, and thetas for ten highest-theta periods found in KIC 5197256.

Period (d)	σ (d)	theta
0.101549	0.000024	3983.94
0.107027	0.000053	1343.2
0.070241	0.000023	1217.91
0.285877	0.000381	994.5
0.126898	0.000075	984.28
0.190713	0.000170	752.41
0.049836	0.000017	526.55
0.090390	0.000057	350.13
0.044873	0.000014	312.97
0.121364	0.000103	312.66

sign was omitted to retain the direction of brightening and dimming in the figures; this is allowable as to determine the amplitude only the difference in magnitude is important). A small representative portion of the resulting light curve (Figure 7) shows approximately one cycle of the dominant period. From this it is estimated that the amplitude is about 1.5 millimags (using the half-amplitude method). As all the other periods are much less significant, no estimation of their amplitudes is given.

Contamination from nearby stars is a common problem among Kepler objects. An inspection of the field using the image collected by Kepler with the coordinates bounding the target against the ALADIN LITE software (Bonnarel et al. 2000) reveals no other point sources included within the pixels Kepler used to collect the data. This strengthens the case that one of the components in the system is a δ Scuti-type variable with several modes of oscillation.

4. Discussion and conclusion

In this study we have presented evidence that KIC 5197256 is an eclipsing binary with at least one pulsating variable. The dominant pulsation period was found to be 0.101549 ± 0.000024 day with an amplitude of ~1.5 millimags. Until future observations indicate otherwise, this object should be included in the class of close binary systems with a δ Scuti component, as studied by Soydugan et al. (2006) and Liakos et al. (2012). If the object is a double-line spectroscopic binary, then the spectroscopic observations planned for a future study should help determine the nature of the system and characteristics of its components.

5. Acknowledgements

This research has made use of "Aladin sky atlas" developed at CDS, Strasbourg Observatory, France, and the SIMBAD database maintained at CDS. The authors would also like to thank an anonymous referee for the helpful suggestions regarding this paper.

References

Association of Universities for Research in Astronomy, Inc. (AURA). 2015, Mikulski Archive for Space Telescopes (MAST; https://archive.stsci.edu/).

Bonnarel, F., et al. 2000, *Astron. Astrophys., Suppl. Ser.*, **143**, 33.

Borucki, W., et al. 2008, in *Exoplanets; Detection, Formation and Dynamics*, eds. Yi-S. Sun, S. F. Mello, and J.-L. Zhou, IAU Symp. 249, Cambridge Univ. Press, Cambridge, 17.

Breger, M. 2000, in *Delta Scuti and Related Stars, Reference Handbook and Proceedings of the 6th Vienna Workshop in Astrophysics*, eds. M. Breger and M. H. Montgomery, ASP Conf. Ser. 210, Astron. Soc. Pacific, San Francisco, 3.

Høg, E., et al. 2000, *Astron. Astrophys.*, **355**, L27.

Kirk, J., et al. 2015, Kepler Eclipsing Binary Stars (Catalog V3; in preparation; http://keplerebs.villanova.edu/).

Liakos, A., Niarchos, P., Soydugan, E., and Zasche, P. 2012, *Mon. Not. Roy. Astron. Soc.*, **422**, 1250.

Lomb, N. R. 1976, *Astrophys. Space Sci.*, **39**, 447.

Scargle, J. D. 1982, *Astrophys. J.*, **263**, 835.

Soydugan, E., et al. 2006, *Mon. Not. Roy Astron. Soc*, **366**, 1289.

Vanmunster, T., 2007, PERANSO period analysis software (http://www.peranso.com).

Early-Time Flux Measurements of SN 2014J Obtained with Small Robotic Telescopes: Extending the AAVSO Light Curve

Björn Poppe
Thorsten Plaggenborg
Universitätssternwarte Oldenburg, Institute of Physics, Carl von Ossietzky University Oldenburg, Germany; address email correspondence to Björn Poppe, bjoern.poppe@uni-oldenburg.de

WeiKang Zheng
Isaac Shivvers
Department of Astronomy, University of California, Berkeley, CA 94720-3411

Koichi Itagaki
Itagaki Astronomical Observatory, Teppo-cho, Yamagata 990-2492, Japan

Alexei V. Filippenko
Department of Astronomy, University of California, Berkeley, CA 94720-3411

Jutta Kunz
Universitätssternwarte Oldenburg, Institute of Physics, Carl von Ossietzky University Oldenburg, Germany

Received December 9, 2014; revised April 14, 2015; accepted May 20, 2015

Abstract In this work, early-time photometry of supernova (SN) 2014J is presented, extending the AAVSO CCD database to prediscovery dates. The applicability of NASA's small robotic MicroObservatory Network telescopes for photometric measurements is evaluated. Prediscovery and postdiscovery photometry of SN 2014J is measured from images taken by two different telescopes of the network, and is compared to measurements from the Katzman Automatic Imaging Telescope and the Itagaki Observatory. In the early light-curve phase (which exhibits stable spectral behavior with constant color indices), these data agree with reasonably high accuracy (better than 0.05 mag around maximum brightness, and 0.15 mag at earlier times). Owing to the changing spectral energy distribution of the SN and the different spectral characteristics of the systems used, differences increase after maximum light. We augment light curves of SN 2014J downloaded from the American Association of Variable Star Observers (AAVSO) online database with these data, and consider the complete brightness evolution of this important Type Ia SN. Furthermore, the first detection presented here (Jan. 15.427, 2014) appears to be one of the earliest observations of SN 2014J yet published, taken less than a day after the SN exploded.

1. Introduction

Supernova (SN) 2014J in the nearby galaxy M82 is widely believed to be one of the most important Type Ia SN explosions of the last few decades, and appears to be the nearest known SN Ia to explode since SN 1972E or, perhaps, Kepler's SN (Foley *et al.* 2014). The object was discovered by S. J. Fossey and his students on 2014 January 21.805 (UT dates are used throughout this paper). A number of prediscovery observations were subsequently reported (Fossey *et al.* 2014; Ma *et al.* 2014; Denisenko *et al.* 2014), and the first-light time was estimated by Zheng *et al.* (2014) to be January 14.75 ± 0.21.

Despite the fact that M82 was regularly monitored every few days, several robotic SN search programs failed to detect SN 2014J automatically (probably owing to the complex host galaxy morphology; see, for instance, Zheng *et al.* 2014). Other early-time observations, including spectroscopy and photometry, have been published by several groups (for example, Goobar *et al.* 2014; Marion *et al.* 2015; Foley *et al.* 2014; Tsvetkov *et al.* 2014). While these studies indicate that SN 2014J was a normal SN Ia, it also exhibited remarkably severe visual extinction and reddening. According to Foley *et al.* (2014), the observations may be explained by a combination of extinction in the M82 interstellar medium and scattering processes in circumstellar material.

We published prediscovery observations of SN 2014J collected with the NASA MicroObservatory Network telescopes (MOs) in the online American Association of Variable Star Observers International Database (AAVSO AID) (unfiltered observations saved under the observer code "USO01"). The MOs monitor M82 as part of the "interesting objects" catalogue with a cadence of 1 day, if weather permits. The MOs consist of a network of robotic telescopes developed by the Harvard-Smithsonian Center for Astrophysics (Sadler *et al.* 2001) to enable world-wide access to astronomical images for students and interested amateurs over the World Wide Web. The telescopes are located and maintained at professional observatories in the United States. They can be controlled remotely, or specific objects can be chosen from a set of predefined suggestions covering examples of the most important astrophysical objects, such as galaxies, star clusters, and nebulae. The predefined suggestions are imaged each night, sometimes even with more than one telescope (as is the case for M82), and all data are archived for at least two weeks in an online accessible database. In this work, we show that these publicly available data can be used to complement scientific campaigns such as light-curve measurements of supernovae and variable stars.

The aim of this paper is therefore twofold: (a) we show that the robotic MOs are able to perform accurate photometric measurements of variable stars, offering the potential to provide valuable data for the AAVSO database or for inclusion in larger survey campaigns; and (b) we use SN 2014J data from the Katzman Automatic Imaging Telescope (KAIT; Filippenko et al. 2001) and the Itagaki Observatory (Japan) along with data from the MOs to construct measurements in the R_c passband, which can be used to extend the AAVSO light curves into the prediscovery time and may also act as baseline data for further analysis and comparisons with other measurements or explosion models.

2. Material and methods

2.1. Telescopes used in this study
2.1.1. MicroObservatory Net (MO)

A detailed technical description of the system is given by Sadler et al. (2001), and so we only describe the most important features of the MOs. The main optical system is composed of Maksutov telescopes with a primary mirror diameter of 15.24 cm (6 inches) and a nominal focal length of approximately 540 mm. The MicroObservatory Net (MO) CCD cameras contain Kodak KAF 1400 chips, with a resolution of 1000 × 1400 pixels, where each pixel is 6.9 μm on a side. Two different fields of view are realized by using different image scales. Either a binning of 2 × 2 pixels is used (1 degree field of view), or only the inner part of the image is analyzed (1/2 degree field of view), resulting in a scale of 5.0 or 2.5 arcsec per pixel, respectively.

The telescopes are equipped with a filter wheel holding several astronomical filters. The images of M82 were taken either with an RGB filter set or a clear filter; we used the latter as an approximation of the unfiltered measurements of SN 2014J from other telescopes. For our study two different telescopes were used, named "Ben" and "Cecilia," both located at the Whipple Observatory, Arizona, USA. All images were taken with an exposure time of 30 s. Both telescopes take daily images of M82. Our earliest detection of SN 2014J was found in an image taken by Cecilia on January 15, about 6 days before the discovery. It is one of the earliest detections published so far.

Overall, more than 170 observations of SN 2014J by the MOs were collected and analyzed for this study, although only the early prediscovery data have been uploaded to the AAVSO database.

2.1.2. Katzman Automatic Imaging Telescope (KAIT)

Early-time photometry of SN 2014J was obtained with KAIT as part of the Lick Observatory Supernova Search (LOSS; Filippenko et al. 2001; Leaman et al. 2011). The system consists of a 0.76-m primary mirror with a focal length of approximately 6.2 m. The camera is a MicroLine77 (Finger Lakes Instrumentation, Lima, New York) with square 24 μm pixels and an image size of 500 × 500 pixels, resulting in a scale of about 0.79 arcsec per pixel.

2.1.3. Itagaki Astronomical Observatory, Japan

Additional early-time data were taken with a 0.5 m telescope by K. Itagaki (Zheng et al. 2014). This telescope has a focal length of approximately 5.0 m, and it uses an N-83E camera from Bitran (Saitama, Japan) with a KODAK KAF-1001E chip (1024 × 1024 pixels; 24 μm per pixel) with a scale of about 1.45 arcsec per pixel.

2.2. Spectral sensitivity

As described by Zheng et al. (2014), the quantum-efficiency curve of the KAIT system reaches half-peak values at ~3800 and 8900 Å, and the Itagaki system has half-peak values at ~4100 and 8900 Å. The half-peak values for the MOs nominally are ~4100 and 7800 Å (Sadler et al. 2001). The slightly lower efficiency of the MOs at longer wavelengths results in a closer approximation to the standard Johnston R-band filter (half-peak values at ~5600 and 7900 Å). In their paper on the first-light time estimates, Zheng et al. (2014) showed that the overall agreement between the KAIT and Itagaki telescopes is better than 0.02 mag over a B–R difference of ~0.9 mag, with a systematic offset of 0.02 mag. We expect the differences between the MOs and the other telescopes to be slightly larger.

2.3. Data analysis
2.3.1. MicroObservatories for photometric evaluations

For an initial verification of the MO system, catalogued stars of all available images were analyzed with ASTROMETRICA (Version 4.1.0.293; Raab 2014). Astrometric reduction and photometry were done with the help of the UCAC4 star catalogue R-band filter data. ASTROMETRICA automatically determines the photometric zeropoint M_{sys} through a combination of a Gaussian fit and an aperture-photometry procedure. First, to calculate the nominal positions and to identify reference stars, a two-dimensional Gaussian function with a constant offset parameter is fitted as a point-spread function (PSF) for all detected sources. Second, the flux of each star is calculated through aperture photometry, using the constant offset parameter of the Gaussian fit as an estimate of the local sky background (window size of the fit equal to the size of the used aperture). The photometric zeropoint is determined by minimizing the mean deviation of all calculated magnitudes to the catalogued reference stars.

For each detected star the brightness estimate M_{est} is then given by

$$M_{est} = -2.5 \log(I) + M_{sys}, \qquad (1)$$

where I is the background-corrected signal within a predefined circular aperture (width defined by the user; for this study r = 3 pixels).

As a check of the image quality, for all detected catalogue stars the measured positions and brightness estimates were compared against the calculated ones; results are presented in the next section.

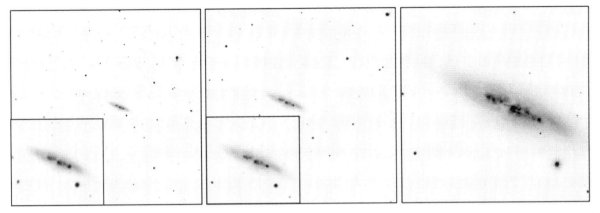

Figure 1. Typical images taken by the telescopes used in this study (all from January 23, 2014). From left to right: An image from "Ben," representative of the MOs; an image from the Itagaki Observatory; and an image from KAIT. Inserts: enlargement of M82 for better visibility of SN 2014J.

2.3.2. Photometry of SN 2014J

In each image I_{SN}, the signal content of SN 2014J was again obtained by the method described in the previous section. Background light from the underlying host galaxy is included in the estimate for the local sky background (as above), and is automatically subtracted. Using the signal content of a photometric comparison star I_{comp} and its catalogued magnitude value $M_{comp,cat}$, the magnitude of SN 2014J is then calculated from

$$M_{SN} = -2.5 \log(I_{SN}) - 2.5(\log I_{comp}) - M_{comp,cat}. \quad (2)$$

This method is equal to the one suggested by the AAVSO (Henden 2015).

A major issue in the analysis of the data is the contribution of the host-galaxy background to the photometric data. As described above, ASTROMETRICA uses the constant offset value determined in the Gaussian-fit procedure to determine the local background. Using this procedure, uncertainties in the photometric analysis will especially become more dominant for the early phase of the SN near the detection limit. Therefore, for the earliest data, as an additional step, independent aperture-based photometry (r = 3 pixels) was performed to verify the evaluations done with ASTROMETRICA. To consider the background from the host galaxy, we followed the method described by Zheng et al. (2014), subtracting a template of the host galaxy obtained from several observations well before the explosion (January 10–13). Aperture photometry of the SN from these galaxy-subtracted images was then calibrated against the magnitude of the reference star with the same aperture-based method described above.

2.3.3. Data merging

When the signal-to-noise ratio (SNR) is high, the dominant source of differences between two sets of stars imaged with different systems is the spectral sensitivity variations among the detectors (for example, Riess et al. 1999). If data from different telescopes are to be compared or merged into one light curve, these variations in the spectral sensitivity of the systems must be taken into account, especially in the case of unfiltered measurements as used in this study. However, as long as the spectral properties of the observed object remain constant, the differences between two systems can be approximated by an additional "intersystemic" magnitude correction constant (see also Zheng et al. 2014). As shown below, the deviations between the intersystemic differences are comparably low and will be considered in the uncertainties of the measurements.

2.3.4. Construction and complementation of light curves in the AAVSO International Database

As the unfiltered passbands from the four telescopes used in this study are relatively similar to standard R passbands, the unfiltered data are calibrated by relative photometry to the r' passband of the UCAC4 catalogue (Zacharias et al. 2013). AAVSO R-band curves are usually given in the Cousins R_c passband. Owing to the small field of view of the KAIT system, no comparison stars could be found in the AAVSO comparison star catalogue, so a direct calibration to r' was not possible. On the other hand, catalogued reference stars for both catalogues are available in the MO images. Therefore, observations from all four systems are first transformed to the r' passband and merged into a single light curve. In a second step, the transformation to the R_c passband is performed for the merged light curve of all four systems simultaneously by choosing the reference star from the MO images. As the light curves from the AAVSO database contain few prediscovery data, this work aims to estimate the early-time light curve of SN 2014J in the R_c passband, thereby complementing the existing dataset.

3. Results

3.1. Photometric utility of the NASA MicroObservatories

Figure 1 shows typical images from the telescopes used for this study. As observations from two identical MOs were used, only one representative image is shown in the figure. It is clear that increasing focal lengths reveal progressively more structure in the host galaxy M82. As the scientific suitability of the Itagaki and KAIT systems have already been well established (Zheng et al. 2014), here we will focus on the MOs. Despite the different focal lengths of the systems, SN 2014J is clearly visible in all images.

To show the applicability of the MOs for accurate photometry, we analyzed the field stars of the images (Figure 1, left panel) in more detail. Altogether we use 176 observations taken between January 15 and March 10 using two different telescopes named "Ben" (105 observations) and "Cecilia" (71 observations). Typically around 80 UCAC4 reference stars were detectable in each image. Astrometric reductions yielded a mean position determination better than 0.3". Instrumental zeropoint calculations were performed with ASTROMETRICA by finding the value resulting in the smallest mean deviation between the measured and catalogued brightness values. The mean deviations for all reference stars was found to be 0.07 ± 0.04 mag for Ben and 0.07 ± 0.02 mag for Cecilia over a color index range of B–V = 0.14 mag through 1.18 mag. Such good agreement over this broad range shows that the unfiltered data can reasonably be compared directly to the r'-band values of the UCAC4 catalogue after applying an appropriate linear offset. Within the brightness range of the reference stars (8–14 mag), no instrumental deviations from linearity indicating saturation or underexposure were observed.

The field-star analysis described above can be influenced from day to day by variable stars, which may have an impact on the mean instrumental magnitude. Therefore, this global analysis of all stars acted only as a first evaluation. In the next step, brightness measurements of four nonvariable comparison stars from the AAVSO Variable Star Plotter (AAVSO 2015; chart number 13184CVH) in the vicinity of SN 2014J were used to evaluate the stability and accuracy of the MOs more specifically. The brightness values from the AAVSO comparison star catalogue are given in the Cousins R_c passband, whereas UCAC4 uses the Sloan r' catalogue. As described above, in the first step the corresponding r' values are used for relative photometry.

To detect undesired instrumental nonlinearity effects caused by underexposure or overexposure of the CCD, we include stars both brighter and fainter than SN 2014J. Stars near SN 2014J were preferentially chosen to minimize and detect effects such as sudden cloud cover or imaging problems. Observations showing deviations from the catalogued values of the reference stars larger than 0.1 mag were visually inspected for weather conditions, and if obvious cloud cover was detectable they were excluded from the study.

3.2. Relative photometry in the UCAC4 r' passband

For the KAIT and Itagaki telescopes, relative photometry in the r' band was performed with the same reference star as used by Zheng et al. (2014) (J2000 coordinates R.A. $09^h 55^m 46.1^s$, Dec. +69° 42' 01.8", USNO-B1.0 R2 = 15.09 mag). However, to be comparable with the MOs the instrumental magnitudes have been calibrated to the reference values of the UCAC4 catalogue in the r' passband (15.49 mag). As the reference star was not visible on the MO images, for both Ben and Cecilia the brightness values were calibrated against the r' of AAVSO comparison star 000-BLG-317 (Table 1). As shown above, the unfiltered, R, and r' passbands are all quite similar, and assuming a simple linear offset between observations through different passbands introduces only moderate errors. Figure 2 illustrates the results of this procedure, as well as the data obtained by aperture photometry on galaxy-subtracted images in the first few days.

As Figure 2 shows, the final reduced data agree quite well, with characteristic differences less than 0.05 mag near peak

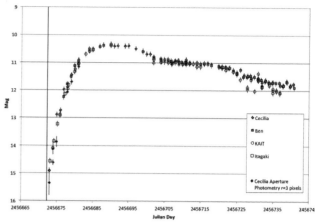

Figure 2. Brightness measurements of the four systems used after calibration to the r' passband. The vertical line indicates the first-light time estimate by Zheng et al. (2014).

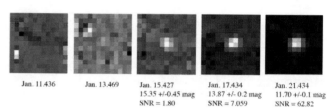

Figure 3. Development of SN 2014J in M82 after subtraction of the host galaxy (all images taken with Cecilia; the nominal detection limit at SNR = 3.0 is 14.8 mag). The first image in which SN 2014J is visible was taken on Jan. 15.427.

brightness. In the early phase of the light curve the scatter is around 0.15 mag. Probably owing to spectral changes of the SN and the different spectral sensitivities, the agreement between the systems gets worse around the second maximum. As we seek only to obtain complementary values at early phases, we neglect observations after the first maximum.

3.3. Analysis of the earliest SN detection

Our earliest detection of SN 2014J was found in an image from Cecilia (January 15.427, JD 2456672.928). Figure 3 shows the development of the SN after subtraction of the host galaxy. As can be seen, the images reveal a first detection on January 15 (SNR = 1.80) with 15.35 ± 0.45 mag. Though the SNR is low, and therefore the measurement uncertainty is large, the SN is clearly detected; thus, we include the data point in the light curve, and it appears to match later (high-SNR) data well. Fitting a Gaussian PSF and a linear background galaxy light contribution, we obtain a brightness of 14.91 ± 0.2 mag; the two methods thus agree within their uncertainties.

Additionally, the values obtained by the two different photometric evaluations agree well, and they fit to the slightly later data taken by the Itagaki Observatory (see Figure 2).

3.4. AAVSO International Database light curve evaluation

Figure 4a shows the mean values of the data submitted by various observers to the AAVSO International Database in the BVR_c bandpasses (AAVSO 2014). Averaging was performed by taking the mean values of all submitted data within a time bin of 1 day. Owing to the high reddening within the host galaxy M82, the difference between the V and B bands is quite large. From the AAVSO data the maximum was estimated by a second-order

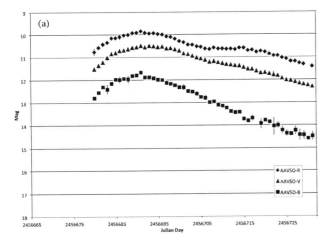

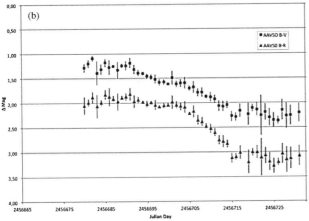

Figure 4. Light curve and color-index development of SN 2014J according to the AAVSO database. (a) Mean brightness and 1σ deviations in B, V, and R_c. (b) Mean color indices.

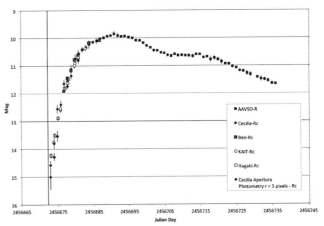

Figure 5. AAVSO light curve complemented by data obtained from the four telescopes used in this study.

polynomial fit to be JD 2456690.5 ± 0.25. Zheng et al. (2014) estimate the first light of SN 2014J to be JD 2456672.25 ± 0.21; the rise time to the B-band maximum is therefore 18.25 ± 0.32 days. According to the AAVSO data, the uncorrected decline parameter $\Delta m_{15}(B)$ in the B band is approximately 1.13 ± 0.15 mag. The date for the second maximum can be estimated for the R band to be 25 ± 0.5 days from B-band maximum, or at JD 2456715 ± 0.5. After the second maximum, the supernova starts to fade slowly, and it declines more rapidly in B than in the other passbands.

As Figure 4b indicates, the color index B–V is around 1.25 ± 0.2 mag before or near maximum brightness, followed by the usual decline after the maximum phase.

3.5. Transformation to the R_c passband

Finally, to compare and complement the AAVSO International Database with our own measurements, we used the tabulated difference between r' and R_c observations of the reference star 000-BLG-317 ($\Delta_{r'-Rc}$ = –0.34 mag (Henden et al. 2014); Table 1) to transform all measurements to R_c. As described previously, because all passbands used in this project are similar, the assumption of a simple linear offset between observations through different passbands introduces only small errors. The result is shown in Figure 5. As can be seen in the overlap region, the agreement between the AAVSO data and our measurements is generally good.

4. Discussion

4.1. MicroObservatories

The analysis of the data herein from two MOs show that these small robotic telescopes with standard imaging parameters are well suited for astrometric and photometric measurements. Unfiltered photometry over a wide brightness range (8–14 mag) and a broad color range (B–V index between 0.14 and 1.18 mag) mesh well with UCAC4 data taken through the r' passband. Moreover, as the mean astrometric deviation is around 0.3", the accuracy requirements of the Minor Planet Center are fulfilled (better than 1"), making these devices useful for astrometry as well (IAU 2014). As another example of the scientific utility of the MOs, Fowler (2013) used MO data to determine the phase diagram of the eclipsing binary star system CQ Cep to high accuracy (better than 0.004 day, though high photometric precision was not needed for that study).

Table 1. Analysis of reference stars for relative photometry of SN 2014J in the UCAC4 r' passband. The value $m_{UCAC4,r'}$ is the catalogued value of the brightness, and Δm characterizes the mean deviations and standard deviations for all observations (105 for Ben, 71 for Cecilia). For 000-BLG-317 the value from the AAVSO comparison star catalogue (R_c) is tabulated. It is used to transform the data from r' to R_c.

Reference Star	R.A. (J2000) h m s	Dec. (J2000) ° ' "	$m_{UCAC4,r'}$	Δm Ben	Δm Cecilia
000-BLG-312	09 55 33	+69 38 55	9.883	0.04 ± 0.05	0.03 ± 0.03
000-BLG-314	09 54 14	+69 37 47	10.943	0.03 ± 0.05	0.04 ± 0.03
000-BLG-317	09 56 33	+69 39 17	12.138 (R_c = 11.798)	0.03 ± 0.05	0.02 ± 0.02
000-BLG-321	09 56 38	+68 41 18	13.314	0.04 ± 0.05	0.04 ± 0.04

Because the MO telescopes offer only clear or RGB filters, their supernova science applications are restricted to first-light detections or relative photometry in the absence of strong spectral changes. As SN 2014J fulfills this latter criterion in the pre-maximum phase, we were able to complement the brightness curves of the AAVSO.

4.2. SN 2014J

From the AAVSO International Database, the B-band maximum was estimated to be at JD 2456690.5 ± 0.25. According to the data, the decline parameter is $\Delta m_{15}(B) = 1.13 \pm 0.15$ mag. The date for the second maximum can be estimated from the R-band data to be 25 ± 0.5 days after B maximum, or at JD 2456715 ± 0.5.

Maximum B-passband times were determined by Marion et al. (2015) to be JD 2456690.2 ± 0.13 (February 1.74 ± 0.13) and by Tsvetkov et al. (2014) to be JD 2456691.4 (February 2.9), similar to our results. The same studies derived Δm_{15} values in the B band between 1.11 ± 0.02 (Marion et al. 2015) and 1.01 (Tsvetkov et al. 2014). Our value measured from the AAVSO database is consistent with the spanned interval. However, for a more detailed analysis and classification of SN 2014J (which is beyond the scope of this paper), additional corrections to Δm_{15}, accounting for extinction and host-galaxy reddening, must be included (see Phillips et al. 1999 or Leibundgut 2000 for an overview).

As Figure 4b indicates, the color index B–V is around 1.25 ± 0.2 mag before and near maximum brightness, increasing to approximately 2.25 around JD 2456670. Again, the value agrees well with the data obtained by Tsvetkov et al. (2014; approximately 1.3 ± 0.1 mag before maximum brightness and 2.3 mag at JD 2456670).

According to Leibundgut (2000), the typical absorption-corrected B–V value is −0.07 ± 0.3 mag. The remarkably high value measured for SN 2014J is undoubtedly caused by extreme dust reddening. Goobar et al. (2014) estimate a host-galaxy reddening for M82 of $E(B-V)_{host} \approx 1.2$ mag, while Schlafly and Finkbeiner (2011) report a Milky Way dust reddening of $E(B-V)_{MW} \approx 0.14$ mag toward M82. Adoption of these values yields a corrected index of B–V ≈ −0.09 mag, in good agreement with that of typical Type Ia supernovae. Foley et al. (2014) give a more detailed model based on Hubble Space Telescope ultraviolet spectra and multi-wavelength observations.

The color index appears to be nearly constant between first detection and B-band maximum, and then moves redward, thus representing the usual expected development – normal Type Ia supernovae show little B–V color change until maximum light, and then exhibit a change of −1.1 mag over the 30 days past maximum (for example, Ford et al. 1993; Leibundgut 2000).

According to Ganeshalingam et al. (2011), the rise times of spectroscopically normal Type Ia supernovae are generally 18.03 ± 0.24 days in B, while those exhibiting high-velocity spectral features have shorter rise times of 16.63 ± 0.29 days. Additionally, normal Type Ia supernovae exhibit a Δm_{15} value in the B band of approximately 1.1 (Phillips et al. 1993). In agreement with other studies (for example, Goobar et al. 2014; Tsvetkov et al. 2014), our derived values for Δm_{15} (1.13 ± 0.15 mag) and the rise time of 18.25 days are consistent with those of spectroscopically normal Type Ia supernovae. Although SN 2014J may be classified as a normal SN Ia, some high-velocity spectral features have been measured in its spectrum (Goobar et al. 2014), and some parameters are near the boundaries between the normal and high-velocity subclasses (Marion et al. 2015).

The MOs monitored M82 routinely each night around the same time. In this work an early-time unfiltered light curve was derived, complementing and supporting the data of Zheng et al. (2014).

The first image in which the SN was detected was obtained on January 15.427, slightly earlier than the observation on January 15.5705 presented by Zheng et al. (2014). The detected brightness value agrees well with their measurements and supports their estimated first-light time (January 14.75 ± 0.21), and it is in good agreement with the very early observations published by Goobar et al. (2014).

5. Conclusion

NASA's MicroObservatories can be used for photometric and astrometric observations with accuracy sufficient for some studies. Early-time photometry of SN 2014J is compared to measurements performed with the KAIT and Itagaki systems. The measurements of all systems were merged and transformed to a single photometric system so as to complement the AAVSO online database with prediscovery observations of SN 2014J. Typical parameters used to characterize Type Ia supernovae were derived from the AAVSO database, and they agree well with values from other studies.

6. Acknowledgements

The authors thank H. Raab and A. Henden for fruitful discussions. A.V.F.'s group and KAIT received financial assistance from the TABASGO Foundation, the Sylvia and Jim Katzman Foundation, the Christopher R. Redlich Fund, the Richard and Rhoda Goldman Fund, and National Science Foundation grant AST–1211916. This research was made possible through the use of the AAVSO Photometric All-Sky Survey (APASS), funded by the Robert Martin Ayers Sciences Fund. We acknowledge with thanks the variable star observations from the AAVSO International Database contributed by observers worldwide and used in this research.

References

AAVSO. 2014, observations from the AAVSO International Database (http://www.aavso.org).

AAVSO. 2015, Variable Star Plotter (VSP; http://www.aavso.org/vsp).

Denisenko, D., et al. 2014, *Astron. Telegram*, No. 5795, 1.

Filippenko, A. V., Li, W. D., Treffers, R. R., and Modjaz, M. 2001, in *Small-Telescope Astronomy on Global Scales*, ASP Conf. Ser. 246, B. Paczyński, W.-P. Chen, and C. Lemme, eds. Astronomical Society of the Pacific, San Francisco, 121.

Foley, R. J., et al. 2014, *Mon. Not. Roy. Astron. Soc.*, **443**, 2887.

Ford, C. H., Herbst, W., Richmond, M. W., Baker, M. L., Filippenko, A. V., Treffers, R. R., Paik, Y., and Benson, P. J. 1993, *Astron. J.*, **106**, 1101.

Fossey, J., Cooke, B., Pollack, G., Wilde, M., and Wright, T. 2014, *Cent. Bur. Electron. Telegrams*, No. 3792, 1.

Fowler, M. J. F. 2013, *J. Br. Astron. Assoc.*, **123**, 49.

Ganeshalingam, M., Li, W., and Filippenko, A. V. 2011, *Mon. Not. Roy. Astron. Soc.*, **416**, 2607.

Goobar, A., et al. 2014, *Astrophys. J.*, **784**, L12.

Henden, A. A., et al. 2014, AAVSO Photometric All-Sky Survey, data release 8 (http://www.aavso.org/apass).

Henden, A., et al. 2015, AAVSO Guide to CCD Photometry, (http://www.aavso.org/sites/default/files/publications_files/ccd_photometry_guide/CCDPhotometryGuide.pdf).

IAU-Minor Planet Center. 2014, "Guide to Minor Body Astrometry," 2014 (http://www.minorplanetcenter.net/iau/info/Astrometry.html).

Leaman, J., Li, W., Chornock, R., and Filippenko, A. V. 2011, *Mon. Not. Roy. Astron. Soc.*, **412**, 1419.

Leibundgut, B. 2000, *Astron. Astrophys. Rev.*, **10**, 179.

Ma, B., Wei, P., Shang, Z., Wang, L., and Wang, X. 2014, *Astron. Telegram*, No. 5794, 1.

Marion, G. H., et al. 2015, *Astrophys. J.*, **798**, 39.

Phillips, M. M. 1993, *Astrophys. J.*, **413**, L105.

Phillips, M. M., Lira, P., Suntzeff, N. B., Schommer, R. A., Hamuy, M., and Maza, J. 1999, *Astron. J.*, **118**, 1766.

Raab, H. 2014, ASTROMETRICA (Version 4.1.0.293).

Riess, A. G., et al. 1999, *Astron. J.*, **118**, 2675.

Sadler, P., et al. 2001, *J. Sci. Education Technol.*, **10**, 39.

Schlafly, E. F., and Finkbeiner, D. P. 2011, *Astrophys. J.*, **737**, 103.

Tsvetkov, D. Y., Metlov, V. G., Shugarov, S. Y., Tarasova, T. N., Pavlyuk, N. N. 2014, *Contrib. Astron. Obs. Skalnaté Pleso*, 44, 67.

Zacharias, N., Finch, C. T., Girard, T. M., Henden, A., Bartlett, J. L., Monet, D. G., and Zacharias, M. I. 2013, *Astron. J.*, **145**, 44.

Zheng, W., et al. 2014, *Astrophys. J.*, **783**, L24.

Recently Determined Light Elements for the δ Scuti Star ZZ Microscopii

Roy Andrew Axelsen
P.O. Box 706, Kenmore, Queensland 4069, Australia; address email correspondence to R. Axelsen, reaxelsen@gmail.com

Tim Napier-Munn
49 Limosa Street, Bellbowrie, Queensland 4070, Australia

Received February 18, 2015; revised May 19, 2015; accepted May 28, 2015

Abstract The δ Scuti star ZZ Microscopii (HD 199757) was studied by photoelectric photometry (PEP) on three nights in 2008 and by DSLR photometry on three nights in 2014. PEP yielded 51 magnitude measurements in V, including 4 peaks of the light curve, and DSLR photometry yielded 622 measurements, including 14 peaks of the light curve. Fourier analysis of the DSLR photometric data found a principle frequency F1 of 14.8853 (0.0001) c/d, and a harmonic frequency 2F1 of 29.7706 (0.0007) c/d, similar to the results of others. Another frequency F2 of 22.2049 (0.0025) c/d, of much lower amplitude than F1, was identified. F2 is higher than the frequency (19.15 c/d) previously reported in the literature, and its accuracy is regarded as uncertain as the semi-amplitude of F2 is low. Regression analysis of an O–C diagram, plotted from 33 historical times of maximum from 1960 to 2003, 4 times of maximum from our PEP in 2008, and 14 times of maximum light from our DSLR photometry in 2014 indicated that a cubic regression provided the best fit. The fitted curve confirms conclusions of others that the period of ZZ Mic was increasing at a constant rate during the years 1960 to 2003, and indicates that the period has decreased during more recent years. The following cubic ephemeris was derived, with zero epoch defined as the first peak of the DSLR photometry light curve on 19 July 2014: T_{max} (HJD) = 2456858.0131 (0.0002) – 7.644 (2.532) × 10^{-19} E^3 – 2.646 (0.973) × 10^{-13} E^2 + 0.06717917 (0.00000001) E.

1. Introduction

ZZ Mic (HD 699757) is a δ Scuti star whose variability was first reported by Churms and Evans (1961). It has a very short principal period of 0.0672 d (Churms and Evans 1961; Leung 1968; Chambliss 1971; Derekas et al. 2009; Kim and Moon 2009). The amplitude of this pulsation is approximately 0.35 mag (Chambliss 1971; Balona and Martin 1978; Derekas et al. 2009). Percy (1976, as quoted by Kim and Moon 2009) analyzed the data published by Leung (1968), and reported two periods of 0.0654 d and 0.0513 d. Derekas (2009) determined the second period to be 0.0522 d, but with a much smaller amplitude (approximately 0.03 mag) than that of the principal period.

Kim and Moon (2009) published a study which dealt mainly with astrophysical properties of ZZ Mic, but included analysis of an O–C diagram of their own data and data by others collected between September 1960 and July 2003, a span of nearly 43 years. These authors fitted a second order polynomial expression to the O–C data. The fitted curve was concave up, indicating that the period of the star was increasing at a constant rate.

Since more than 11 years have elapsed since the time of the most recent (2003) data in Kim and Moon's (2009) publication, it is considered timely to report further studies of the periods of ZZ Mic, incorporating data from photoelectric and DSLR photometry.

2. Data and analysis

Photoelectric photometry was performed in 2008 using an SSP-5 photometer from Optec Inc, Lowell, Michigan. The instrument was fitted with a Hamamatsu R6358 multialkali photomultiplier tube, and Johnson V and B photometric filters. Measurements were taken through a Celestron C9.25 Schmidt-Cassegrain telescope, on a Losmandy GM-8 German equatorial mount. The comparison and check stars were HD 200027 and HD 200320, respectively. Non-transformed magnitudes in V were calculated, since the color indices of the variable and comparison stars are relatively close. Photometry was performed on three nights in 2008, namely, 31 July, 9 August, and 28 October. A total of 51 magnitude determinations were made over a period of 5 hours 20 minutes.

DSLR photometry was performed on RAW images taken with a Canon EOS 500D DSLR camera through a refracting telescope with an aperture of 80 mm at f/7.5, mounted on a Losmandy GM-8 German equatorial mount. Images were obtained on three nights in 2014, namely, 19, 27, and 28 July. A total of 622 magnitude determinations were made over a total observing period (including meridian flips) of 25 hours 35 minutes.

Photometric data reduction from DSLR instrumental magnitudes utilized the software package AIP4WIN (Berry and Burnel 2011). The comparison and check stars were the same as those chosen for photoelectric photometry. Transformed magnitudes in V were calculated using transformation coefficients for the blue and green channels of the DSLR sensor, calculated from images of standard stars in the E regions (Menzies et al. 1989).

The time of each magnitude measurement (the mid point of each set of three PEP measurements, and the mid point of each DSLR exposure) was recorded initially in Julian Days (JD), and subsequently converted to Heliocentric Julian Days (HJD). The heliocentric correction was calculated for the mid point in time of the observation set for each night, and that correction was applied to each time in JD for the corresponding night.

Fourier analysis used the software package PERIOD04. The software package PERANSO was used to determine the time of maximum light for each peak in the light curve, calculated as the maximum value of a 6th order polynomial expression fitted to each peak.

3. Results

Examples of light curves of ZZ Mic from photoelectric and DSLR photometry are shown in Figures 1 and 2, respectively.

The results of Fourier analysis (Table 1 and Figure 3) identified a principal frequency F1 14.8853 (0.0001) c/d which corresponds to a period of 0.0672 d, confirming the results of others (Churms and Evans 1961; Leung 1968; Chambliss 1971; Derekas *et al.* 2009; Kim and Moon 2009). The harmonic frequency 2F1, 29.7706 (0.0007) c/d is close to that in the literature, but the other identified frequency, 22.2049 (0.0025) c/d, is higher than the frequency of 19.15 c/d previously identified (Derekas *et al.* 2009).

An O–C diagram was drawn from the data in Table 3 of Kim and Moon (2009), combined with our own data. Kim and Moon stated that 34 times of maximum were utilized, but only 33 are tabulated in their paper, comprising their own observations, as well as those of Churms and Evans (1961), Leung (1968), Chambliss (1971), and Balona and Martin (1978). The publications representing the sources of the times of maximum are quoted by Kim and Moon, but are not referenced to each individual data point. Therefore, an attempt has been made to do this retrospectively. Table 2 lists the times of maximum for the O–C calculations from Kim and Moon's (2009) paper (rows 1 to 33 of Table 2), and includes the references to the sources of the data as interpreted by us from information in that paper. Table 2 also includes our own data (rows 34 to 51), comprising 4 times of maximum from photoelectric photometry in 2008 and 14 times of maximum from DSLR photometry in 2014. Figure 4 illustrates the O–C diagram, and the cubic (third order polynomial) model fitted to the data. The cubic regression for the O–C data is given by the equation:

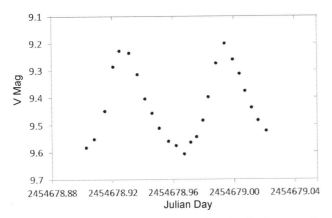

Figure 1. Light curve of ZZ Mic obtained by photoelectric photometry from observations taken on one night over 2 hours 51 minutes.

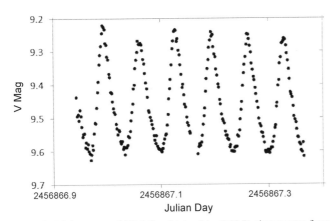

Figure 2. Light curve of ZZ Mic obtained by DSLR photometry from observations taken on one night over 10 hours 10 minutes.

Table 1. Results of Fourier analysis by the software package PERIOD04. The second frequency (2F1, 19.7706 c/d) is almost exactly twice the principal frequency (F1), and is therefore a harmonic of F1. The values of sigma were calculated in PERIOD04 using a Monte Carlo simulation with 100 processes, in which frequency and phase uncertainties were not uncoupled. Note that the semi amplitude of F2 is substantially less than that of F1.

F	Frequency (c/d)	Frequency Sigma	Semi-Amplitude	Semi-Amplitude Sigma
F1	14.8853	0.0001	0.164	0.001
2F1	29.7706	0.0007	0.038	0.001
F2	22.2049	0.0025	0.009	0.001

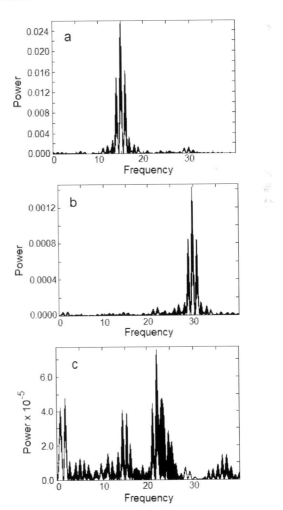

Figure 3. Fourier analysis by PERIOD04 of DSLR photometric data of ZZ Mic. Power spectra show: (a) the principal frequency F1 at 14.8853 c/d, (b) a harmonic frequency 2F1 at 29.7706 c/d, and (c) another frequency F2 at 22.2049 c/d, with the latter having a substantially lower amplitude than F1.

Table 2. Data for ZZ Mic from which the O–C diagram in Figure 2 was drawn.

Maximum	TOM (HJD)	Epoch	O–C	Source*
1	2437192.31400	0	0.001000	1
2	2437249.28190	848	0.000955	1
3	2439320.21610	31675	0.002573	2
4	2439321.22350	31690	0.002286	2
5	2439330.15630	31823	0.000255	2
6	2439330.22340	31824	0.000176	2
7	2439331.22870	31839	–0.002212	2
8	2440449.56250	48486	–0.000221	3
9	2440449.63030	48487	0.000399	3
10	2440450.50300	48500	–0.000230	3
11	2440450.57040	48501	–0.000009	3
12	2440450.63730	48502	–0.000288	3
13	2440451.51050	48515	–0.000418	3
14	2440451.57790	48516	–0.000197	3
15	2440451.64630	48517	0.001024	3
16	2443356.33840	91755	–0.000261	4
17	2443356.40640	91756	0.000560	4
18	2449996.66450	190600	–0.000208	5
19	2449997.60620	190614	0.000983	5
20	2449997.60670	190614	0.001483	5
21	2449997.67360	190615	0.001204	5
22	2449997.67370	190615	0.001304	5
23	2449998.61390	190629	0.000996	5
24	2450405.58600	196687	0.001623	5
25	2450406.59410	196702	0.002036	5
26	2452237.29430	223953	0.002401	5
27	2452474.57070	227485	0.001938	5
28	2452477.52550	227529	0.000854	5
29	2452493.51640	227767	0.003109	5
30	2452495.53240	227797	0.003734	5
31	2452496.40690	227810	0.004904	5
32	2452496.47110	227811	0.001925	5
33	2452842.51120	232962	0.002069	5
34	2454678.92288	260298	0.003681	6
35	2454678.98960	260299	0.003229	6
36	2454687.99196	260433	0.003575	6
37	2454767.93562	261623	0.004012	6
38	2456858.01271	292735	0.002457	7
39	2456858.08010	292736	0.002663	7
40	2456865.94133	292853	0.003925	7
41	2456866.00668	292854	0.002104	7
42	2456866.07460	292855	0.002837	7
43	2456866.14110	292856	0.002163	7
44	2456866.20880	292857	0.002680	7
45	2456866.27554	292858	0.002244	7
46	2456866.94769	292868	0.002597	7
47	2456867.01531	292869	0.003043	7
48	2456867.08185	292870	0.002402	7
49	2456867.14919	292871	0.002568	7
50	2456867.21692	292872	0.003116	7
51	2456867.28342	292873	0.002434	7

*Notes: Rows 1 to 33 represent the data in Table 3 of Kim and Moon (2009). The sources of the data, in the last column on the right (as interpreted by us, from information in Kim and Moon's paper), are (1) Churms and Evans 1961; (2) Leung 1968; (3) Chambliss 1971; (4) Balona and Martin 1978; (5) Kim and Moon 2009; (6) photoelectric photometric data of the present authors; (7) DSLR photometric data of the present authors. Calculation of the O–C values is based on the ephemeris used by Kim and Moon (2009) for their own calculations, namely, T_0 (HJD) 2437192.313 and period 0.0671786 d.

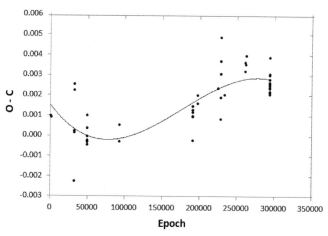

Figure 4. O–C diagram of ZZ Mic incorporating data from the literature and the authors' PEP and DSLR photometry. The diagram spans the years 1960 to 2014. The fitted curve represents a cubic (third order polynomial) expression. The earlier part of the curve, which is concave up, confirms that the period was increasing until 2003 approximately. In more recent years, the period has decreased.

$$O-C = 0.0015\ (0.0006) - 7.64\ (2.53) \times 10^{-19}\ E^3 + 4.07\ (1.28) \times 10^{-13}\ E^2 - 4.91\ (1.82) \times 10^{-8}\ E \quad (1)$$

The fit was found to be superior to those of linear or second order polynomial models (see discussion in section 4 below). A cubic ephemeris was therefore calculated for the behavior of ZZ Mic, and represents new light elements for the star:

$$T_{max}\ (HJD) = 2456858.0131\ (0.0002) - 7.644\ (2.532) \times 10^{-19}\ E^3 - 2.646\ (0.973) \times 10^{-13}\ E^2 + 0.06717917\ (0.00000001)\ E \quad (2)$$

The zero epoch in this ephemeris is the time of the first maximum in the set of DSLR observations from 19 July 2014. The period of the star on that date is given in the last term in Equation 2, 0.06717917 (0.00000001) d.

4. Discussion

The principal frequency from Fourier analysis of the present DSLR photometric observations, 14.8853 c/d, corresponds to a period of 0.0672 d, a result similar to that reported in several previous publications (Churms and Evans 1961; Leung 1968; Chambliss 1971; Derekas et al. 2006; Kim and Moon 2009), and essentially identical to the more precise value of 0.06717917 d in a cubic ephemeris calculated by us from times of maximum light published by others and combined with our own data.

The Fourier harmonic 2F1 reported herein is mentioned in the literature only by Derekas et al. (2009). The additional frequency F2 of 22.2049 c/d corresponds to a period of 0.04505 d, which is shorter than the period of 0.0513 d reported by Percy (1976, as quoted by Kim and Moon 2009) and the period of 0.0522 d found by Derekas (2006). The period ratio of 0.78 was reported by both Percy (1976, as quoted by Kim and Moon 2009) and Derekas (2009) and falls within the range (0.74–0.78) accepted for the ratio of the first overtone to the fundamental frequency for a δ Scuti star pulsating in the radial mode (Breger 1979). In contrast, the ratio obtained by us is only 0.67, which falls outside the accepted range for such stars. It is therefore considered that the accuracy of this ratio is uncertain, as the semi-amplitude of F2 is low (0.009 magnitude in V) and substantially less than that of F1 at 0.164 magnitude (Table 1).

Analysis of the O–C diagram was undertaken by performing linear, second order, and third order polynomial regression analysis of the 51 data points available. The three models were compared by "extra sum of squares" analysis to determined whether each polynomial provided any statistically significant improvement in fit to its predecessor, based on the residual sum of squares of the fit. It was found that the linear and quadratic models were not statistically different from each other ($P = 0.26$). However, the cubic model was superior to both the linear and quadratic models ($P < 0.01$). All its four coefficients, including the intercept, were statistically significant ($P < 0.01$) and the adjusted coefficient of determination, R^2_{adj}, was greater than that of the other two (0.593 compared with 0.524 and 0.521); as the models are nested the values of R^2_{adj} are directly comparable. We therefore believe that the cubic model of the O–C data is superior to the others and is the model of choice in this case.

Weighted regression was attempted using the reciprocal of the variance of O–C values obtained at similar epochs as weights. Interesting results were obtained, including modestly different coefficients and improved residuals. This approach is promising but requires more analysis and has not been pursued further in the present work.

Therefore, the behavior of ZZ Mic, for the data available, is best described by a cubic ephemeris. The most recent previous O–C analysis of ZZ Mic in the literature is the paper by Kim and Moon (2009) who analyzed observations made across nearly 43 years, between 1960 and 2003. Those authors found that the period of the star was increasing at a constant rate across the time covered by that data set. The conclusion from the more recent data obtained by us is that the period of the star has decreased since the last 2003 data obtained by Kim and Moon.

5. Acknowledgements

The photoelectric photometer used in 2008 to obtain part of the data reported in this paper was purchased with the award to one of us (RAA) of an Edward Corbould Research Fund grant from the Astronomical Association of Queensland, Brisbane, Australia.

References

Balona, L. A., and Martin, W. L. 1978, *Mon. Not. Roy. Astron. Soc.*, **184**, 11.

Berry, R., and Burnell, J. 2011, "Astronomical Image Processing for Windows," version 2.4.0, provided with The Handbook of Astronomical Image Processing, Willmann-Bell, Richmond, VA.

Breger, M. 1979, *Publ. Astron. Soc. Pacific*, **91**, 5.

Chambliss, C. R. 1971, *Astrophys. J.*, **165**, 365.

Churms, J., and Evans, D. S. 1961, *Observatory*, **81**, 25.

Derekas, A., et al. 2009, *Mon. Not. Roy. Astron. Soc.*, **394**, 995.

Kim, C., and Moon, B.-K. 2009, *Publ. Astron. Soc. Pacific*, **121**, 478.

Leung, K.-C. 1968, *Astron. J.*, **73**, 500.

Menzies, J. W., Cousins, A. W. J., Banfield, R. M., and Laing, J. D. 1989, *S. Afr. Astron. Obs. Circ.*, **13**, 1.

Percy, J. R. 1976 (quoted by Kim and Moon, 2009), in *Proceedings of the Solar and Stellar Pulsation Conference, Los Alamos Scientific Laboratory*. Eds. Cox, A. N. and Deupree, R. G., Los Alamos Scientific Laboratory, Los Alamos, NM, 61.

A Photometric Study of ASAS J184708-3340.2: an Eclipsing Binary with Total Eclipses

Robert C. Berrington
Ball State University, Dept. of Physics and Astronomy, Muncie, IN 47306; rberring@bsu.edu

Erin M. Tuhey
Ball State University, Dept. of Physics and Astronomy, Muncie, IN 47306; emtuhey@bsu.edu

Received July 22, 2014; revised August 20, 2014 and May 16, 2015; accepted May 26, 2015

Abstract We present new multi-band differential aperture photometry of the eclipsing variable star ASAS J184708-3340.2. The light curves are analyzed with the Wilson-Devinney model to determine best-fit stellar models. Our models show that ASAS J184708-3340.2 is consistent with an overcontact eclipsing binary (W Ursae Majoris) system with total eclipses.

1. Introduction

The All Sky Automated Survey (Pojmański 1997; ASAS) catalog entry star 184708-3340.2 (R.A. $18^h 47^m 8^s$, Dec. $-33° 40' 12"$ (J2000.0)) was originally classified as an eclipsing contact binary system (EC) with a period of $P = 0.28174$ day by Pojmański and Maciejewski (2005). The ASAS is a large-area photometric sky survey covering the entire southern hemisphere and a portion of the northern hemisphere ($\delta < 28°$) (Pojmański 1997). The survey provides continuous photometric and simultaneous monitoring of bright sources ($V \lesssim 14$ and $I \lesssim 13$) in both the V and I bands (Pojmański 2002). Currently, only the Johnson V band data have been made available to the public (http://www.astrouw.edu.pl/asas/). A convenient web-based interface accessing the classification database (Pojmański 2002, 2003; Pojmański and Maciejewski 2004, 2005; Pojmański et al. 2005) of all large-amplitude variable stars southward of declination $-28°$ is provided, and reports a 13.27 V-band magnitude and V-band variability amplitude of 0.68 magnitude for ASAS J184708-3340.2. Because of the large area coverage of the survey, temporal resolution of the V and I band photometry is not fine enough to calculate times of minima for these short-period systems, and makes a detailed photometric study of short-period variable systems difficult.

In this paper we present a new extensive photometric study of ASAS J184708-3340.2. The paper is organized as follows. Observational data acquisition and reduction methods are presented in section 2. Time analysis of the photometric light curve and Wilson-Devinney (WD) models is presented in section 3. Discussion of the results and conclusions is presented in section 4.

2. Observational data

We present new three-filter differential aperture photometry of the eclipsing variable star ASAS J184708-3340.2. The data were taken by the SARA-South 0.6-meter telescope located at the Cerro Tololo Inter-American Observatory (CTIO). All exposures were acquired by the Astronomical Research Cameras (ARC), Inc. camera, which uses the thinned, back-illuminated E2V CCD42-40 CCD that contains a 2k × 2k array of 13.5µm pixels, and were taken through the Johnson-Cousins B, V, and R (R_c) filters on the nights of June 22, 2013 (JD 2456465), and July 4, 2013 (JD 2456477), with 2 × 2 on-chip binning to expedite chip readout times. All images were bias and dark current subtracted, and flat field corrected using the CCDRED reduction package found in the Image Reduction and Analysis Facility (IRAF; distributed by the National Optical Astronomy Observatories (http://iraf.net/), version 2.16. All photometry presented is differential aperture photometry and was performed on the target eclipsing candidate and two comparison standards by the AIP4WIN (v2.2.0) photometry package (Berry and Burnell 2005). Over the two nights a total of 154 were images were taken in B, 155 images in V, and 157 images in R. Figure 1 shows a digitized sky survey POSS2/UKSTU red image (https://archive.stsci.edu/cgi-bin/dss_form) with the eclipsing star candidate and

Figure 1. Star field containing the variable star ASAS J184708-3340.2. The location of the variable star is shown along with the comparison (C1) star and the check (C2) star used to calculate the differential magnitudes reported in Figure 2.

the two comparison stars marked, with the primary comparison star labeled C1 and the secondary comparison (check) star labeled C2. The folded light curves (see section 3.1) for the instrumental differential B, V, and R_c magnitudes are shown in Figure 2, and are defined as the variable star magnitude minus C1 (Variable – C1). Also shown (bottom panel) in Figure 2 is the differential V magnitude of C1 minus C2. The comparison light curve was inspected for variability. None was found.

Because of the magnitude range of ASAS J184708-3340.2, the required exposure lengths necessary to obtain the targeted signal-to-noise ratio for the variable meant the stars with accurately measured Johnson-Cousins B, V, and R_c magnitudes were saturated. To solve this issue, we took several exposures with shortened exposure times that did not saturate the brighter stars with accurately known magnitudes. These exposures allowed us to calibrate the magnitudes for C1 and C2. The stars used for this calibration were the Tycho stars TYC 7412-1069-1, TYC 7412-1794-1, and TYC 7412-1940-1 (Høg et al. 2000).

Measured instrumental B and V differential magnitudes for C1 and the aforementioned Tycho stars were converted to Johnson B and V magnitudes by comparison with known calibrated magnitudes of the Tycho stars. The calibrated magnitudes were averaged together after individual values were found to be consistent with each other ($<1.5\sigma$) in both the Johnson B and V bands. The star C1 was found to have measured Johnson B and V magnitude, of 13.90 ± 0.10 and 12.77 ± 0.12, respectively. The calibrated V light curve with the (B–V) color index versus orbital phase is shown in Figure 3. Error bars have been removed for clarity. The orbital phase (Φ) is defined as:

$$\Phi = \frac{T-T_0}{P} - \mathrm{Int}\left(\frac{T-T_0}{P}\right), \quad (1)$$

where T_0 is the ephemeris epoch and is the time of minimum of a primary eclipse. Throughout this paper we will use the value of 2456465.595055 for T_0. The variable T is the time of observation, and the period of the orbit is given by P. The value of Φ typically ranges from a minimum of 0 to a maximum of 1.0. We can also define negative orbital phase values by the relation $\Phi - 1$. All light curve figures will plot phase values (–0.6, 0.6). Simultaneous B and V magnitudes are used to determine (B–V) colors by linear interpolation between measured B magnitudes.

3. Analysis

3.1. Period analysis and ephemerides

Heliocentric Julian dates (HJD) for the observed times of minimum were calculated for each of the B-, V-, and R_c-band light curves shown in Figure 2 for all observed primary and secondary minima. A total of one primary eclipse and two secondary eclipses were observed for each band. The times of minimum were determined by the algorithm described by Kwee and van Woerden (1956). Times of minimum from

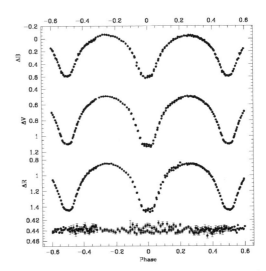

Figure 2. Folded light curves for differential aperture Johnson-Cousins B-, V-, and R_c-band magnitudes. Phase values are defined by Equation 1. Top three panels show the folded light curves for Johnson B (top panel), Johnson V (middle panel), and Cousins R_c (bottom panel) magnitudes. Bottom panel shows differential Johnson V-band magnitudes for the comparison minus the check star. All error bars are 1σ error bars, and for the top three panels are smaller than the point size. Repeated points do not show error bars (points outside the phase range of (–0.5, 0.5)).

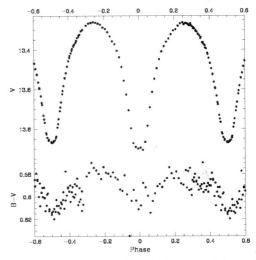

Figure 3. Folded light curve for differential aperture Johnson V band magnitudes (top panel) and (B–V) color (bottom panel) versus orbital phase. Phase values are defined by Equation 1. Error bars are not shown for clarity. All (B–V) colors are calculated by subtracting linearly interpolated B magnitudes from measured V magnitudes. Magnitudes calibration is discussed in section 2.

differing band passes were compared and no significant offsets or wavelength dependent trends were observed. Times of minimum from each of the band passes were averaged together and reported in Table 1 along with 1σ error bars.

Light curves were inspected using the PERANSO (v2.51) software (CBA Belgium Observatory 2011) to determine the orbital period by applying the analysis of variance (ANOVA) statistic which uses periodic orthogonal polynomials to fit observed light curves (Schwarzenberg-Czerny 1996). Our best-fit orbital period was found to be 0.28179 ± 0.00020 day and is consistent ($<1\sigma$) with the orbital period reported by Pojmański and Maciejewski (2005). The resulting linear ephemeris becomes:

$$T_{min} = 2456465.59506(22) + 0.28179(24) \, E. \quad (2)$$

where the variable E represents the epoch number, and is a count of orbital periods from the epoch $T_0 = 2456465.59506$. Figure 2 shows the folded differential magnitudes versus orbital phase for ASAS J184708-3340.2 for the B, V, and R_C Johnson-Cousins bands folded over the period determined by the current photometric study.

The observed minus calculated residual times of minimum $(O-C)$ were determined from Equation 2 and are given in Table 1 along with 1σ error bars. The best-fit linear line determined by a linear regression to the $(O-C)$ residual values is shown in Figure 4, and indicates a small correction to the period to a value of 0.2817412 ± 0.0000008 day. The newly determined period is consistent with our previously determined value from PERANSO at $<1\sigma$ deviation. For all subsequent fitting the period of 0.2817412 ± 0.0000008 day will be assumed. The use of either period has no impact on the conclusions of this study.

3.2. Effective temperature and spectral type

Effective temperature and spectral type are estimated from the $(B-V)$ color index values measured at orbital quadrature $(\Phi = \pm 0.25)$ with a value of $(B-V) = 0.58 \pm 0.16$. The interstellar extinction estimate following Schlafly and Finkbeiner (2011) at the galactic coordinates for the object is $E(B-V) = 0.14$. The resulting intrinsic color becomes $(B-V)_0 = 0.44 \pm 0.16$. Effective temperatures and errors were estimated by Table 3 from Flower (1996) to be $T_{eff} = 6541 \pm 700$K. The corresponding stellar spectral type is F5 (Fitzgerald 1970).

3.3. Light curve analysis

All observations taken during this study were analyzed using the Physics of Eclipsing Binaries (PHOEBE) software package (v0.31a) (Prša and Zwitter 2005). The PHOEBE software package is a modeling package that provides a convenient, intuitive graphical user interface (GUI) to the WD code (Wilson and Devinney 1971).

All three Johnson-Cousins B, V, and R_C bands were fit simultaneously by the following procedure. Initial fits were performed assuming a common convective envelope in direct thermal contact, resulting in a common surface temperature of $T_{eff} = 6541$K determined by the procedure discussed in section 3.1. Orbital period was set to the value of 0.2817412 day and not allowed to vary. Surface temperatures imply that the outer envelopes are convective, so the gravity brightening coefficients B_1 and B_2, defined by the flux dependency $F \propto g^\beta$, were initially set at the common value consistent with a convective envelope of 0.32 (Lucy 1967). The more recent studies of Alencar and Vaz (1997) and Alencar et al. (1999) predict values for $\beta \approx 0.4$. We initially set the standard stellar bolometric albedo $A_1 = A_2 = 0.5$ as suggested by Rucinski (1969) with two possible reflections.

The fitting procedure was used to determine the best-fit stellar models and orbital parameters from the observed light curves shown in Figure 2. Initial fits were performed assuming a common convective envelope in thermal contact, which assumes similar surface temperatures for both.

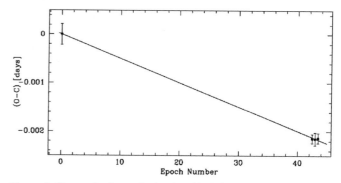

Figure 4. Observed minus calculated residual times of minimum $(O-C)$ versus orbital epoch number. All point values are given in Table 1. Secondary times of minimum are plotted at half integer values, and all error bars are 1σ error bars. Solid curve shows the best-fit linear line determined by a linear regression fit to the $(O-C)$ residual values.

Table 1. Calculated heliocentric Julian dates (HJD) for the observed times of minimum of ASAS J184708-3340.2.

T_{min}	Eclipse	E	$(O-C)$
$2456465.59506 \pm 0.000218$	p	0	0 ± 0.000218
$2456477.56899 \pm 0.000094$	s	42.5	-0.002145 ± 0.000094
$2456477.70988 \pm 0.000132$	p	43	-0.00215 ± 0.000132
$2456477.85079 \pm 0.000103$	s	43.5	-0.002135 ± 0.000103

Notes: Calculated heliocentric Julian dates (HJD) for the observed times of minimum (column 1) with the type of minima (column 2). Observed minus Calculated (O-C) residual (column 4) values are given for the linear ephemeris given in Equation 2. All reported times are averaged from the individual B-, V-, and R-band times of minimum determined by the algorithm descried by Kwee and van Woerden (1956). All (O-C) values are given in units of days with primary eclipse values determined from integral epoch numbers, and secondary eclipse values determined from half integral epoch numbers (column 3).

After normalization of the stellar luminosity, the light curve was crudely fit by altering the stellar shape by fitting the Kopal (Ω) parameter. The Kopal parameter describes the equipotential surface that the stars fill. This defines the shape of the stars, and strongly influences the global morphology of the light curve.

After the fit could no longer be improved, we started to consider the other parameters to fit the light curve. These parameters included the effective temperature of the secondary star $T_{eff,2}$, the mass ratio $q = M_2 / M_1$, and the orbital inclination i, and we varied them to improve the overall model fit. Minor improvement of the of the best-fit model could be achieved by decoupling stellar luminosities from T_{eff}. We interpreted this as the possibility that the stars could have differing surface temperatures. All further model fits were performed assuming the primary and secondary components might have differing surface temperatures, but did not include decoupling stellar luminosities from T_{eff}.

All model fits were performed with a limb darkening correction. PHOEBE allows for differing functional forms to be specified by the user. Late-type stars ($T_{eff} < 9000$K) are best described by the logarithmic law which was first suggested by Klinglesmith and Sobieski (1970), and later supported by the more recent studies of Diaz-Cordoves and Gimenez (1992) and van Hamme (1993). For all of our model fits, we assumed a logarithmic limb darkening law. The values for the linear (x_λ) and non-linear (y_λ) coefficients were determined at each fitting iteration by the van Hamme (1993) interpolation tables.

Figures 5 through 7 show the folded Johnson B-, Johnson V-, and Cousins R_C-band light curves along with the synthetic light curve calculated by the best-fit model, respectively. The best-fit models were determined by the aforementioned fitting procedure. The parameters along with 1σ error bars describing this best-fit model are given in column 3 of Table 2.

The filling factor is defined by the inner and outer critical equipotential surfaces that pass through the L_1 and L_2 Lagrangian points of the system, and is given by the following equation:

$$\mathcal{F} = \frac{\Omega(L_1) - \Omega}{\Omega(L_1) - \Omega(L_2)} \qquad (3)$$

where Ω is the equipotential surface describing the stellar surface, and $\Omega(L_1)$ and $\Omega(L_2)$ are the equipotential surfaces that pass through the Lagrangian points L_1 and L_2, respectively. For our system the these equipotential surfaces are $\Omega(L_1)$ = 5.958, and $\Omega(L_2)$ = 5.348. The best-fit model is consistent with an overcontact binary described by a filling factor $\mathcal{F} = 0.151$, and consistent with an overcontact binary system. Graphical representations for the best-fit WD model is shown in Figure 8.

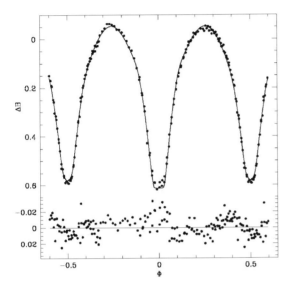

Figure 5. Best-fit WD model fit (solid curve) to the folded light curve for differential Johnson B-band magnitudes (top panel). The best-fit orbital parameters used to determine the light curve model are given in Table 2. The bottom panel shows residuals from the best-fit model (solid curve). Error bars are omitted from the points for clarity.

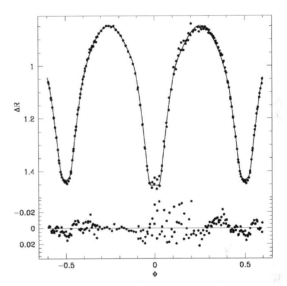

Figure 7. Best-fit WD model fit (solid curve) to the folded light curve for differential R_C-band magnitudes (top panel). The best-fit orbital parameters used to determine the light curve model are given in Table 2. The bottom panel shows residuals from the best-fit model (solid curve). Error bars are omitted from the points for clarity.

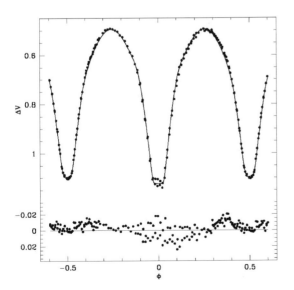

Figure 6. Best-fit WD model fit (solid curve) to the folded light curve for differential Johnson V-band magnitudes (top panel). The best-fit orbital parameters used to determine the light curve model are given in Table 2. The bottom panel shows residuals from the best-fit model (solid curve). Error bars are omitted from the points for clarity.

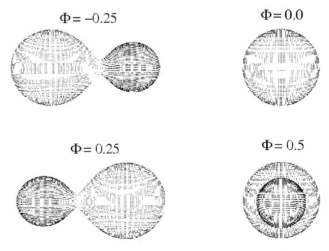

Figure 8. Graphical representation for the best-fit WD model. Orbital phase for each panel is given in the upper right corner. The best-fit orbital parameters used to determine the light curve model are given in Table 2.

Table 2. Model parameters for ASAS J184708-3340.2 determined by the best-fit WD model.

Parameter	Symbol	Value
Period	P [days]	0.2817412 ± 0.0000008
Epoch	T_0 [HJD]	2456465.59506 ± 0.0000008
Inclination	i [°]	88.1 ± 0.2
Surface Temperature	$T_{eff,1}$ [K]	6541 ± 700
	$T_{eff,2}$ [K]	6305 ± 700
Surface Potential	$\Omega_{1,2}$ [—]	5.866 ± 0.003
Mass Ratio	q [—]	2.51 ± 0.01
Luminosity	$[L_1/(L_1+L_2)]B$	0.347 ± 0.001
	$[L_1/(L_1+L_2)]V$	0.338 ± 0.001
	$[L_1/(L_1+L_2)]R_C$	0.329 ± 0.001
Limb Darkening	$x_{bol,1,2}$	0.64
	$y_{bol,1,2}$	0.24
	$x_{B,1,2}$	0.80
	$y_{B,1,2}$	0.25
	$x_{V,1,2}$	0.71
	$y_{V,1,2}$	0.28
	$x_{R,1,2}$	0.62
	$y_{R,1,2}$	0.28

Notes: Values for each parameter (column 1) along with brief descriptions (column 2) that specify the best-fit WD stellar model are given in column 3. Some parameters can be further specified by a the numerical value 1 for the primary stellar component, or 2 for the secondary stellar component. Fitting procedure is described in section 3.3. Surface potentials for both stars for contact/overcontact binaries is defined to be of equal value for both stars. Errors for surface temperatures (T_{eff}) were estimated from color values in Figure 3. All remaining errors are 1σ errors. Please note that the parameters L_1 and L_2 refer to the luminosities of primary and secondary components, respectively.

4. Discussion and conclusions

Given the parameters in Table 2, we can estimate the distance to ASAS J184708-3340.2. Rucinski and Duerbeck (1997) determined that the absolute visual magnitude is given by

$$M_V = -4.44 \log_{10}(P) + 3.02 (B-V)_0 + 0.12 \qquad (4)$$

to within an accuracy of ±0.22. The distance modulus of the system $(m - M) = 9.38$ for the value obtained from Equation 4 after accounting for the extinction ($A_V = 0.44$) determined from the color excess given in section 3.1. This corresponds to a distance of 613 pc.

This study has shown that ASAS J184708-3340.8 is well described as a W Ursae Majoris overcontact binary with a filling factor $\mathcal{F} = 0.151$, and both eclipses passing through totality with an inclination of $i = 88.1° ± 0.2$. Surface temperatures for the stellar components do differ, and possibly indicate that the system is in poor thermal contact. Additional spectroscopic follow-up will be necessary to place further constraints on the spectral types, stellar masses, and orbital velocities to better improve our knowledge of the system.

References

Alencar, S. H. P., and Vaz, L. P. R. 1997, *Astron. Astrophys.*, **326**, 257.

Alencar, S. H. P., Vaz, L. P. R., and Nordlund, Å. 1999, *Astron. Astrophys.*, **346**, 556.

Berry, R., and Burnell, J. 2005, *Handbook of Astronomical Image Processing*, Willmann-Bell, Richmond.

CBA Belgium Observatory. 2011, PERANSO (v2.51) software, Flanders, Belgium (http://www.cbabelgium.com/).

Diaz-Cordoves, J., and Gimenez, A. 1992, *Astron. Astrophys.*, **259**, 227.

Fitzgerald, M. P. 1970, *Astron. Astrophys.*, **4**, 234.

Flower, P. J. 1996, *Astrophys. J.*, **469**, 355.

Høg, E., et al. 2000, *Astron. Astrophys.*, **355**, L27.

Klinglesmith, D. A., and Sobieski, S. 1970, *Astron. J.*, **75**, 175.

Kwee, K. K., and van Woerden, H. 1956, *Bull. Astron. Inst. Netherlands*, **12**, 327.

Lucy, L. B. 1967, *Z. Astrophys.*, **65**, 89.

Pojmański, G. 1997, *Acta Astron.*, **47**, 467.

Pojmański, G. 2002, *Acta Astron.*, **52**, 397.

Pojmański, G. 2003, *Acta Astron.*, **53**, 341.

Pojmański, G., and Maciejewski, G. 2004, *Acta Astron.*, **54**, 153.

Pojmański, G., and Maciejewski, G. 2005, *Acta Astron.*, **55**, 97.

Pojmański, G., Pilecki, B., and Szczygiel, D. 2005, *Acta Astron.*, **55**, 275.

Prša, A., and Zwitter, T. 2005, *Astrophys. J.*, **628**, 426 (PHOEBE software package v0.31a).

Rucinski, S. M. 1969, *Acta Astron.*, **19**, 245.

Rucinski, S. M., and Duerbeck, H. W. 1997, *Publ. Astron. Soc. Pacific*, **109**, 1340.

Schlafly, E. F., and Finkbeiner, D. P. 2011, *Astrophys. J.*, **737**, 103.

Schwarzenberg-Czerny, A. 1996, *Astrophys. J., Lett. Ed.*, **460**, L107.

van Hamme, W. 1993, *Astron. J.*, **106**, 2096.

Wilson, R. E., and Devinney, E. J. 1971, *Astrophys. J.*, **166**, 605.

A Binary Model for the Emission Line Star FX Velorum

Mel Blake
Maisey Hunter
Department of Physics and Earth Science, University of North Alabama, One Harrison Plaza, Florence, AL 35630; blake@una.edu

Received January 22, 2014; revised June 23, 2014, February 26, 2015; accepted May 26, 2015

Abstract FX Vel is a southern, young variable star which shows large variations in brightness. In examining the environments where planets may form, disks around young stars provide important clues as to how long such disks might last. We discuss several possible scenarios for the structure of FX Vel, including a binary model similar to EE Cep and the possibility that FX Vel may be a UXor variable, a class of young stars with disks. This could also explain the colors and irregular variation in brightness of the star. We suggest FX Vel may be a blue straggler member of the open cluster ASCC48 based upon its position on the cluster CMD and proper motions. More data is required examine the alternate possibility of FX Vel being a member of Teutsch 101. A spectroscopic study of FX Vel would be valuable.

1. Introduction

The formation of planetary systems in different environments is of great importance in informing planet searches of which spectral types are suited to harbor planets. Most searches target dwarfs and it is commonly thought that the short lifetimes of the most massive stars may not allow time for planets to form around them. However, binary systems exist which consist of a B-star and a less massive star where a disk seems to be present around the less massive companion, such as EE Cep (Gałan et al. 2012; Mikołajewski et al. 2005). In the case of EE Cep, the binary consists of a Be star with a K-type companion. The disk which is responsible for the deep changes in brightness is associated with the lower-mass companion star, rather than the primary, which eclipses the Be star about every five years. The disk around the secondary precesses, changing the contact angle of the eclipse with the disk of the primary. This causes the depth of the eclipse to vary in depth and for the eclipse itself to have a characteristic asymmetric shape (for example, Gałan et al. 2012; Graczyk et al. 2003). There seems to be a central region that has been cleared out in the disk, causing a short term increase in brightness as the gap passes over the primary.

One reason for addressing this issue is the short lifetime of the Be stars. The Be stars are highly variable optically and have gaseous disks formed around a rapidly-rotating B-type star (Rivinius et al. 2013). These stars should last less than a few tens of millions of years before expiring. The current state of the disk around the companions can therefore be compared to models of the disks with these constraints. In addition, the Be stars should not have strayed far from the region they formed, which may include a star forming region or star cluster. As a result, finding the Be star in a cluster should be able to produce an age for the binary, which in turn provides a timescale for the creation of the structures in the disk of the companion star. The secondary star should survive the explosion of the primary, and it is of great interest to know if planets will form and what their fate would be. Identifying analogs of EE Cep for comparison will help address this issue.

FX Vel came to our attention during a search of the All-Sky Automated Survey for variable stars (Pojmański 2004) in the direction of open star clusters. Using the WEBDA Open Cluster Database (Masaryk University 2015) we searched within the radius of open clusters for variables. FX Vel appears in the direction of the cluster ASCC48, identified by Kharchenko et al. (2005), although it does lie near the edge of the cluster. The star is bright (V = 9.77), making it accessible to small telescopes. The large, varying eclipse depth and the presence of a massive star suggested possible similarities to EE Cep. We therefore decided to investigate the star further.

FX Vel was first identified as an emission line star by Merrill and Burwell (1950), who did an objective prism survey. FX Vel is their star No. 202 and they give the Hα line as strong. The variable star nature of the star was found in the study of Strohmeier et al. (1968), who gave an amplitude 1.20 photographic magnitudes. They classify the star as EW or EB and comment that the secondary eclipses were very deep. They provide a period of 1.052565 days for the system and give ten times of minimum, but do not provide light curves. Heinze (1976) included FX Vel in his catalogue of emission line stars, where it is his star 174. The spectrum is described as having moderately sharp Hα, with a strength lying between moderate and weak compared to the continuum. No note was made about any Hα variability.

Eggen (1978) studied 60 contact systems using photographic observations in the Stromgren and Hβ system. He obtained a spectral type of B9 III-IV for FX Vel and an orbital period of 1.052565 days, based upon a dozen observations. His results did not match those of previous results, showing possibly more than a magnitude-deep primary eclipse. He comments that the Hα observations of Wray (1966) indicate that the emission is weak in FX Vel, which differs from the Strohmeier et al. (1968) results. This may suggest Hα variability in FX Vel. Thé et al. (1994) include FX Vel as a Herbig Ae/Be star, while Friedemann et al. (1996) include FX Vel in their catalogue of binaries with associated IRAS sources. In particular, they include FX Vel as one of their group A stars, which have IR fluxes that vary as λ^{-1}, rather than λ^{-2} as expected from a purely stellar spectrum. They conclude the presence of circumstellar dust around these types of stars. Tisserand et al. (2013) give a spectral type of A3III for FX Vel.

Most recently, Miroshnichenko et al. (2007) investigated FS CMa stars and included FX Vel as a member of this group.

FS CMa stars are B[e] stars which are also associated with dusty disks, many of which have lower-mass companions. One characteristic of the FS CMa stars is the unusual infrared emission in their spectra which suggests on-going formation of dust. These stars also tend to lie in the field, away from sites of current star formation. Miroshnichenko *et al.* (2007) point out the contradictory studies of the amplitude of variability of FX Vel, and note that their examination of the ASAS light curve rules out the short 1.05-day orbital period found by Strohmeier (1968) and Eggen (1978), since the minimum which occurred around JD 2452000 lasted for more than two months. Miroshnichenko *et al.* (2007) suggest the irregular light curve is similar to other isolated Ae/Be stars except for the lack of reddening. The spectra of Miroshnichenko *et al.* (2007) were taken in December 2004 and show double-peaked profiles in their Balmer lines with stronger blue-ward emission. They suggest this may indicate perturbations in the disk around the Be star.

2. Cluster membership

The original reason for FX Vel drawing our attention was that it is within the search radius of the cluster ASCC48 when searched with the WEBDA open clusters database. Kharchenko *et al.* (2005) searched for unidentified open clusters and identified 109 new open cluster candidates. ASCC48 has not been studied since. If FX Vel is associated with a cluster this could provide an age estimate of the star. We note that a cluster this age is likely too old to host a Be star (for example, Tarasov and Malchenko 2012). The possibility is that FX Vel does not contain a Be star, given the spectral type of A3III, as discussed in section 3. Figure 1 shows the color-magnitude diagram of ASCC48, with FX Vel indicated. We plot the isochrones from the Padova group (Bressan *et al.* 2012) using the parameters obtained by Kharchenko *et al.* (2005), log age = 9.1, E(B–V) = 0.00, V – Mv = 8.01, and include the photometry from the study of Kharchenko *et al.* (2005). We include isochrones with Z = 0.019, 0.008, and 0.004, with the Z = 0.019 matching the CMD best. A spectroscopic study to obtain the metallicity of the cluster would be helpful. There is large scatter present, but FX Vel does not appear to be a main sequence member of this cluster although it may be a blue straggler. To investigate this further we compared the proper motion of FX Vel to that of the cluster members. In Figure 2 we show the proper motions μ_α and μ_δ of cluster members with greater than 50% probability of membership based on proper motions (Kharchenko *et al.* 2005), as well as the proper motion of FX Vel from Tycho-2 (Høg *et al.* 2000). FX Vel is close to the region occupied by the probable cluster members and is closer to the cluster of points than others which have a greater than 50% probability of being members. We conclude that if FX Vel is a cluster member, it is most likely a blue straggler. However, many stars that appear to members based upon their photometry are excluded based upon proper motions in the WEBDA database, so a radial velocity study would be useful. We can't conclude whether FX Vel is a member or not.

Another cluster near FX Vel is DSHJ0832.5-3807 (Teutsch 101), identified by Kronberger *et al.* (2006) using 2MASS data.

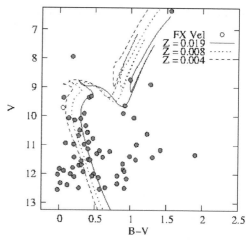

Figure 1. The Color-Magnitude Diagram (CMD) of ASCC 48 with FX Vel. We include the photometry of cluster members from Kharchenko *et al.* (2005) and isochrones from Bressan *et al.* (2012). The best match to the main sequence and the giant branch of the cluster is that for Z = 0.019. FX Vel (open circle) does not appear to be a cluster member unless it is a blue straggler.

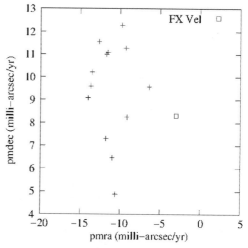

Figure 2. Proper motions of members of ASCC 48 with greater than 50% probability of membership based upon their proper motions (crosses). FX Vel (square) is also plotted and does not appear to be an outlier, implying membership. A radial velocity study would help determine if FX Vel is a member or not.

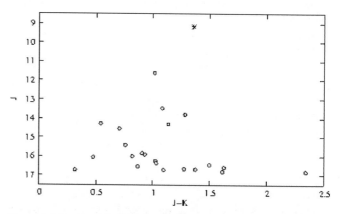

Figure 3. A J versus J-K CMD from the 2MASS database the stars within 1 arminute of the position of Teutsch 101 given in Table 2. The star indicates the position of FX Vel while the circles show data for the stars near the cluster center. FX Vel appears brighter but redder than the candidate cluster stars, but without membership probabilities, reddening, and distances it is difficult to draw any conclusions from this comparison.

Table 1. Data from 2MASS on point sources within the cluster radius of Teusch 101. Null errors are flags given when the magnitude is a 95% confidence upper limit.

2MASS Identification	R.A. (J2000) h m s	Dec. (J2000) ° ′ ″	J	σJ	H	σH	K	σK
08323442-3807212	08 32 34.43	–38 07 21.24	14.310	0.030	13.466	0.033	13.173	0.032
08323204-3807271	08 32 32.05	–38 07 27.13	16.596	0.128	15.742	0.116	14.969	null
08323675-3806253	08 32 36.75	–38 06 25.39	16.730	0.153	16.317	0.188	15.634	null
08323203-3807520	08 32 32.03	–38 07 52.07	16.013	0.072	15.158	0.058	15.192	0.168
08323476-3806519	08 32 34.77	–38 06 51.97	16.540	0.116	16.410	0.195	15.672	0.265
08323091-3807198	08 32 30.92	–38 07 19.89	14.277	0.036	13.883	0.032	13.734	0.054
08323143-3807415	08 32 31.43	–38 07 41.51	16.373	0.111	15.761	0.120	15.342	0.199
08323403-3807013	08 32 34.03	–38 07 01.37	15.912	0.061	15.181	0.063	14.980	0.139
08323492-3807137	08 32 34.92	–38 07 13.77	14.550	0.047	14.060	0.058	13.843	0.060
08323704-3806504	08 32 37.04	–38 06 50.46	16.670	0.152	16.414	0.218	15.392	null
08323204-3807040	08 32 32.04	–38 07 04.06	15.422	0.043	15.009	0.058	14.663	0.114
08323158-3806397	08 32 31.59	–38 06 39.74	16.729	0.148	16.359	0.190	16.408	null
08323250-3807170	08 32 32.50	–38 07 17.01	16.817	0.160	16.144	0.296	14.475	null
08323143-3807065	08 32 31.44	–38 07 06.56	16.063	0.075	15.457	0.079	15.586	0.248
08323331-3806572	08 32 33.31	–38 06 57.23	13.474	0.026	12.688	0.024	12.390	0.021
08323526-3807069	08 32 35.27	–38 07 06.99	16.699	0.155	16.307	0.218	15.323	null
08323497-3807196	08 32 34.97	–38 07 19.63	13.799	0.039	12.891	0.046	12.513	0.030
08323094-3807131	08 32 30.94	–38 07 13.15	16.451	0.112	15.721	0.111	14.954	null
08323305-3807152	08 32 33.05	–38 07 15.29	11.611	0.027	10.865	0.026	10.596	0.021
08323746-3806251	08 32 37.47	–38 06 25.14	16.805	0.154	16.174	0.160	15.193	null
08323369-3806513	08 32 33.70	–38 06 51.40	15.848	0.069	15.268	0.074	14.942	0.142
08323439-3806393	08 32 34.40	–38 06 39.37	16.265	0.094	15.436	0.077	15.241	0.171

We investigate here if FX Vel might be a member of this group. We do not have a reddening, age, or distance estimate or proper motions for this cluster, so we are left with producing a CMD from the stars in the 2MASS database within 1 arcminute of the cluster coordinates given by Kronberger *et al.* (2006). We give the data from the 2MASS database of these stars in Table 1, and the CMD in Figure 3. Interpretation of this plot is difficult with so few stars, but FX Vel is nearly two magnitudes brighter and is redder than the next brightest object which suggests it is too bright to be a member. Without estimates of reddening of each star to correct the colors this conclusion is not very secure so this comparison is inconclusive.

3. The Nature of FX Vel

3.1. Is FX Vel a binary?

As commented on by Miroshnichenko *et al.* (2007), the ASAS (Pojmański *et al.* 2013) has collected several years of data on FX Vel. They rule out the 1.05-day orbital period of the system based on their examination of the length of minimum in the light curve. Since that time, the ASAS has produced more observations of FX Vel, which we consider here. Figure 4 shows the light curve of FX Vel from the ASAS. The striking feature of this dataset is that the dips are varying in both depth and in shape, with considerable variability on each one. This is not easy to explain with a simple eclipse model, and we propose here that what is being observed is the eclipse of the primary star by the precessing disk of a cooler companion. This is similar to the model used to explain the changing shape and depth of the dips in EE Cep.

Figure 5 shows the data for several brightness dips. The first dip in brightness (Figure 5 top left) is asymmetric, with the drop to minimum light taking longer than the subsequent rise to maximum brightness. Interestingly, the dip in brightness is preceded by a shallow drop in brightness. The following dip in

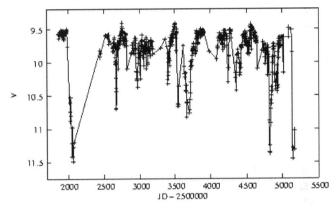

Figure 4. The light curve for FX Vel from the All-Sky Automated Survey (ASAS; Pojmański *et al.* 2013). FX Vel undergoes drops in brightness or more than 1 to 2 magnitudes, with some events deeper than others.

brightness (Figure 5 top right) is a double-dipped event. The first dip in brightness has a slight drop in brightness followed by a very steep drop to minimum light. Unfortunately the following increase in brightness is not covered by the ASAS monitoring. Following this is another dip in brightness, which starts apparently even before maximum brightness is reached. The dip in brightness has several individual fluctuations on a small scale and the slope gets shallower towards minimum light. The brightness first rises sharply and then slowly reaches maximum light. The slow rise has several dips in brightness as well. The third event (Figure 5 bottom left) which we show is poorly measured, but is also preceded by a shallow drop in brightness and appears symmetric. It is very similar to the first. The final event (Figure 5 bottom right) looks similar to the second, but is nearly a magnitude deeper. There is first a very rapid decrease in brightness, followed by an increase in brightness that changes slope. There is then a shallower, less deep drop in brightness after the initial dip in brightness.

The shallower drop in brightness is also double-dipped with a peak in the center of the eclipse.

We hope to understand why there appear to be two types of events for FX Vel, and why some dips in brightness appear to occur in pairs or with shallower events following or preceding them. If the secondary star has a disk around it that is warped by the gravity of the primary the variations in brightness can be understood (Figure 6). The greater the amount of the disk that intercepts light from the primary the greater the drop in brightness and the longer the drop in brightness will be. However, if the disk is precessing, the orientation of the disk can change. In the first case (Figure 6 top) the disk of the primary intersects the disk at three points, creating first a short, shallow brightness drop followed by a longer, deeper eclipse. In the second case (Figure 6 middle) the precession of the disk creates geometry in which the primary intersects the disk at only one location, resulting in a short shallow drop in brightness only (Figure 6 middle). In the final example (Figure 6 bottom) the disk of the primary intersects the disk at three points, but the longer, deeper event precedes the shorter, shallower drop in brightness.

If the primary star has a binary companion, then the secondary star may orbit out of the plane of the disk of the primary, and the secondary's disk will form along the equatorial plane of the secondary star. Miroshnichenko et al. (2007) obtained an Hα spectrum of FX Vel which shows a central dip surrounded by emission peaks with a stronger blue peak than red. Rivinius et al. (2013) suggest that this form of spectrum is observed when the Be star is seen through the plane of the disk from the perspective of Earth. Any difference between the plane of the orbit and that of the disk around the companion will cause precession in the disk of the secondary. As the binaries orbit, the orientation of the disk will be changing from our perspective. A very variable light curve results from these combined changes.

One serious problem with the binary model is that the dips in brightness are very long. If solely caused by orbital motion of the secondary as the disk passes in front of its companion, the disk would seem to be overflowing the Roche lobe of the secondary, or be bigger than the binary system itself. This is difficult to reconcile.

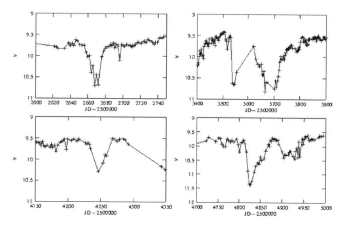

Figure 5. Four eclipse events for FX Vel showing the variety of changes in the eclipses. The first eclipse (top left) is slightly asymmetric, with the drop in brightness being slightly longer than the following increase in brightness after light minimum. The second eclipse (top right) has partial coverage of the first decrease in brightness, but the initial drop in brightness is very rapid. The following drop in brightness is at first rapid and then becomes less shallow near the bottom of the eclipse. The subsequent increase in brightness is rapid followed by a slower increase. The third eclipse (bottom left) has relatively poor coverage, but appears symmetric. The fourth eclipse (bottom right) has a rapid drop in brightness with a slower increase. This is followed by a shallower eclipse that also exhibits an increase in brightness near the bottom.

3.2. Is FX Vel a UX Ori variable?

An alternative explanation has been proposed by S. A. Otero in his remarks on FX Vel in the AAVSO International Variable Star Index (VSX; Watson et al. 2014). The UXor variables are a class of young stars in the pre-main sequence stage of evolution. The stars are emission line A-type stars which exhibit several magnitude drops in brightness (Herbst and Shevchenko 1999) for up to several days or weeks. Dullemond et al. (2003) interpret the brightness variations as being caused by density variations in a disk around the star causing greater or less absorption of the light from the star. This mimics the drops in brightness of FX Vel. In addition FX Vel has colors (Table 2) that resemble the class of UXor variables more so than EE Cep. Tisserand et al. (2013) give a spectral type of A3 IIIe for FX Vel as well, similar to the UXor stars. We include the infrared colors of Be star γ Cas for comparison. Clearly FX Vel has colors more similar to the UXor variables than the Be stars. However, also for comparison, we include the colors of EE Cep, whose light curve resembles that of FX Vel. EE Cep, as we have noted, contains a Be star and a low-mass companion with a disk in which the precession of the disk causes variations in the eclipse depths and lengths. EE Cep has colors that do not match FX Vel well, so its IR colors would seem to argue against an interpretation of FX Vel's light curve variations being caused by a precessing disk around a companion. Observing campaigns would be useful for FX Vel to study variation in Hα and colors during the drops in brightness.

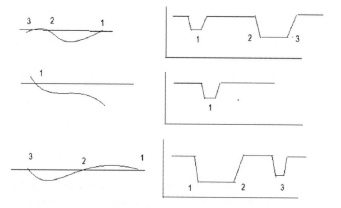

Figure 6. A variety of eclipses can result from a precessing warped disk. In case one (top), the disk of the primary will intersect the disk at three points, resulting in first a short, shallow eclipse followed by a longer, deeper eclipse. In case two (middle), the primary will intersect the disk at only one point, resulting is a single shallow eclipse. In case three (bottom), the primary first intersects the disk creating a longer deep eclipse followed by a short, shallow eclipse. The variety and varying depth of the eclipses seen in Figure 4 is reproduced. The disk may also have a gap in it, and may vary in thickness, making the shapes of the eclipse asymmetric.

Table 2. Infrared Measurements of UXor stars, FX Vel, and γ Cas.

Star	Wavelength/Band (μ)							
	W1 3.35	W2 4.6	W3 11.6	W4 22.1	J 1.25	H 1.65	K 2.17	J–W4
BF Ori	6.84	6.05	3.59	2.20	9.11	8.57	7.90	6.91
SV Cep	6.74	5.98	2.55	0.75	9.35	8.56	7.74	8.60
UX Ori	6.24	5.51	2.92	0.92	8.71	8.04	7.21	7.79
WW Vul	6.58	5.95	3.33	1.56	9.09	8.18	7.28	7.53
XY Per	4.92	3.79	2.55	0.97	7.65	6.92	6.09	6.68
FX Vel	6.53	5.56	2.41	1.05	9.17	8.66	7.81	8.12
γ Cas	1.67	0.45	0.04	-0.30	2.04	1.99	1.76	2.34
EE Cep	7.24	6.70	5.99	5.60	9.60	9.22	8.56	4.00

4. Conclusions and future work

We have examined the star FX Vel and have concluded more data are required to determine if it is a member of the galactic star cluster ASCC 48 based upon its location on the CMD of the cluster and its proper motion. Likewise more data are needed to determine if FX Vel is a member of Teutsch 101. Radial velocity studies of these clusters and FX Vel would be very helpful in resolving this. The light curve of FX Vel has been re-examined and we suggest two possible models to explain the properties of its photometric variability. In one model the changes may be explained by the presence of a precessing disk around the secondary star which intercepts differing amounts of light as each orbit of the secondary occurs. The double dip of one eclipse suggests that the disk may have a warp in it that allows only partial coverage of the primary disk during eclipses at some times. In the second model, supported by the IR colors and length of the eclipses, FX Vel is a UXor variable with a star surrounded by a disk with differing densities. FX Vel has been relatively poorly studied and a great deal could be done to help improve our understanding of this star. We do not have color information for the dips in brightness of the star of the star. We expect that the color should be redder if a disk is causing the large variable brightness changes. Photometric monitoring in multiple bands should be carried out to provide these data. In addition, spectra are needed to study the variation of the spectrum at Hα. Polarization studies before, during, and after eclipses might also help study the properties of the dust.

5. Acknowledgements

We would like to express appreciation to the anonymous referee whose comments and suggestions greatly improved this paper. This publication makes use of data products from the Wide-field Infrared Survey Explorer, which is a joint project of the University of California, Los Angeles, and the Jet Propulsion Laboratory/ California Institute of Technology, funded by the National Aeronautics and Space Administration. This research has made use of the WEBDA database, operated at the Department of Theoretical Physics and Astrophysics of the Masaryk University.

References

Bressan, A., Marigo, P., Girardi, L., Salasnich, B., Dal Cero, C., Rubele, S., and Nanni, A. 2012, *Mon. Not. Roy. Astron. Soc.*, **427**, 127.

Dullemond, C. P., van den Ancker, M. E., Acke, B., and van Boekel, R. 2003, *Astrophys. J.*, **594**, L47.

Eggen, O. J. 1978, *Astron. J.*, **83**, 288.

Friedemann, C., Guertler, J., and Loewe, M. 1996, *Astron. Astrophys., Suppl. Ser.*, **117**, 205.

Gałan, C., et al. 2012, *Astron. Astrophys.*, **544A**, 53.

Graczyk, D., Mikołajewski, M., Tomov, T., Kolev, D., and Iliev, I. 2003, *Astron. Astrophys.*, **403**, 1089.

Heinze, K. G. 1976, *Astrophys. J., Suppl. Ser.*, **30**, 491.

Herbst, W., and Shevchenko, V. S. 1999, *Astron. J.*, **118**, 1043.

Høg, E., et al. 2000, *Astron. Astrophys.*, **355**, L27.

Kharchenko, N. V., Piskunov, A. E., Röser, S., Schilbach, E., and Scholz, R –D. 2005, *Astron. Astrophys.*, **440**, 403.

Kronberger, M., et al. 2006, *Astron. Astrophys.*, **447**, 921.

Masaryk University, Department of Theoretical Physics and Astrophysics. 2015, WEBDA Open Cluster Database (http://www.univie.ac.at/webda/webda.html).

Merrill, P. W., and Burwell, C. G. 1950, *Astrophys. J.*, **112**, 72.

Mikołajewski, M., et al. 2005, *Astrophys. Space Sci.*, **296**, 451.

Miroshnichenko, A. S., et al. 2007, *Astrophys. J.*, **671**, 828.

Pojmański, G. 2004, *Astron. Nachr.*, **325**, 553.

Pojmański, G., Szczygiel, D., and Pilecki, B. 2013, The All-Sky Automated Survey Catalogues (ASAS3; http://www.astrouw.edu.pl/asas/?page=catalogues).

Rivinius, T., Cariofi, A. C., and Martayan, C. 2013, *Astron. Astrophys. Rev.*, **21**, 69.

Strohmeier, W., Ott, H., and Schoffel, E. 1968, *Inf. Bull. Var. Stars*, No. 261, 1.

Tarasov, A. E., and Malchenko, S. L. 2012, *Astron. Lett.*, **38**, 428.

Thé, P. S., de Winter, D., and Pérez, M. R. 1994, *Astron. Astrophys., Suppl. Ser.*, **104**, 315.

Tisserand, P., Clayton, G. C., Welch, D. L., Pilecki, B., Wyrzykowski, L., and Kilkenny, D. 2013, *Astron. Astrophys.*, **551**, 77.

Watson, C., Henden, A. A., and Price, C. A. 2014, AAVSO International Variable Star Index VSX (Watson+, 2006–2014; http://www.aavso.org/vsx; see FX Vel listing, www.aavso.org/vsx/index.php?view=detail.top&oid=37603).

Wray, J. D.,1966, *Astron. J.*, **71**, 403.

Comparison Between Synthetic and Photometric Magnitudes for the Sloan Standard Stars

Anthony Mallama
14012 Lancaster Lane, Bowie, MD 20715; anthony.mallama@gmail.com

Received April 17, 2015; revised May 26, 2015; accepted May 28, 2015

Abstract Synthetic magnitudes derived from STIS/CALSPEC fluxes were compared to photometric magnitudes for standard stars of the Sloan system. The statistics of the magnitude differences are consistent with the stated and intended accuracies of both sources in all five Sloan bands. Close agreement in the Sloan u' band-pass extends magnitude-based verification of STIS/CALSPEC fluxes to near-ultraviolet wavelengths.

1. Introduction

The absolute calibration of spectroscopic and photometric data is critical to many astrophysical investigations. For example, accurate spectral energy distributions (SEDs) are needed for the analysis of light from red-shifted type Ia supernovae when used to study dark energy.

White dwarf stars serve as relatively stable flux standards for spectroscopic and photometric systems. Historically, SEDs of the white dwarfs were determined by Oke (1990) using the 5-meter Hale telescope. The absolute flux of the Sloan magnitude system (Smith *et al.* 2002) is tied to the SEDs from those ground-based observations performed several decades ago. Much progress has been made in the precise radiometric calibration of SEDs in recent years, though. In particular, data from the Space Telescope Imaging Spectrograph (STIS) instrument on-board Hubble are now considered to be the most accurate source of radiometrically calibrated data. The study described in this paper uses the CALSPEC database (Bohlin *et al.* 2014) of STIS SEDs to validate the magnitudes of Sloan primary stars.

Besides verifying the Sloan magnitudes, it is equally important to validate the STIS/CALSPEC SEDs themselves. Bohlin and Landolt (2015) summarize the work that has been performed using visible and near-IR Johnson-Cousins magnitudes as a comparison for CALSPEC fluxes. The present study extends that by using Sloan magnitudes for comparison and also extends it to the near-UV. Thus, this paper assesses the consistency between the absolute radiometric fluxes of the Sloan photometric system and the STIS/CALSPEC database. Since the two systems were developed independently, the comparison may be interpreted as a measure of absolute accuracy.

The methods used in selecting stars for this study and in comparing the CALSPEC fluxes with Sloan magnitudes are described in section 2. Central to this discussion is the equation used for deriving synthetic magnitudes from spectral fluxes. Section 3 lists the synthetic CALSPEC magnitudes along with their differences with respect to the Sloan magnitudes. The statistics of the differences between Sloan and CALSPEC magnitudes are next discussed and an estimate of the consistency between the two systems is given. A similar comparison based on Johnson-Cousins magnitudes (Bohlin and Landolt 2015) is described in section 4. Finally, the conclusions of this study are summarized and a suggestion for follow-up research with small telescopes is offered in section 5.

2. Methods

Stars were chosen for this study by matching the Sloan standards listed by Smith *et al.* (2002) with the CALSPEC standards listed by Bohlin *et al.* (2014). Table 1 lists the six resulting stars and shows that three of them were used as fundamental standards by Smith *et al.* The other stars are ordinary Sloan standards.

Table 1. Sloan and CALSPEC stars.

Star Name	Type*
BD +02 3375	A5
BD +17 4708	sdF8**
BD +21 0607	F2**
BD +26 2606	A5**
BD +29 2091	F5
BD +54 1216	sdF6

*from Bohlin et al. (2014)
**fundamental Sloan standard

The magnitudes used in this study were taken from Table 8 of Smith *et al.* (2002). These values are on the Sloan photometric system and they differ from those of the Sloan survey itself. The magnitudes of Smith *et al.*, which are preferred for photometry, are also available on-line at http://www-star.fnal.gov/ugriz/tab08.dat.

FITS files of CALSPEC data were retrieved from http://www.stsci.edu/hst/observatory/crds/calspec.html. The header of each file was checked to insure SEDs extending over all five Sloan band passes were from the STIS instrument. This insured that the data were of the highest possible quality.

In order to compare CALSPEC and Sloan standards, CALSPEC fluxes were transformed to Sloan magnitudes. These synthetic magnitudes were derived from SEDs by integrating the product of spectral energy multiplied by system response over each Sloan band pass. Equation 1 (Smith *et al*, 2002; Fukugita *et al.* 1996) indicates the relationship among magnitude, flux, and system response.

$$m = -2.5 \frac{\int d(\log \nu)\, f\nu\, S\nu}{\int d(\log \nu)\, S\nu} \quad (1)$$

where m is magnitude, fv is flux, Sv is the system response, and v is frequency. The units are ergs per square centimeter per Hertz per second. The five system response functions referenced by Smith et al. were retrieved from http://www-star.fnal.gov/ugriz/Filters/response.html. Their central wavelengths are u', 355.1; g', 468.6; r', 616.5; i', 748.1; and z', 893.1 nm. After solving for m, a constant of –48.60 was added to place Sloan magnitudes on the absolute AB system (Oke and Gunn 1983; Fukugita et al. 1996).

3. Results

The resulting synthetic magnitudes for the six stars are listed by Sloan band in Table 2. The magnitude differences for each star in the sense "synthetic magnitude minus photometric" are then listed in Table 3. Most of the differences are less than 0.01 magnitude.

Table 2. Synthetic magnitudes from STIS/CALSPEC fluxes.

Star Name	u'	g'	r'	i'	z'
BD +02 3375	11.027	10.132	9.816	9.702	9.672
BD +17 4708	10.569	9.643	9.358	9.269	9.254
BD +21 0607	10.284	9.391	9.117	9.034	9.024
BD +26 2606	10.756	9.884	9.605	9.515	9.496
BD +29 2091	11.356	10.471	10.118	9.997	9.965
BD +54 1216	10.783	9.886	9.589	9.498	9.486

Table 3. Synthetic magnitudes minus photometric magnitudes.

Star Name	u'	g'	r'	i'	z'
BD +02 3375	0.011	0.002	0.007	0.016	0.015
BD +17 4708	0.009	0.003	0.008	0.019	0.024
BD +21 0607	-0.005	-0.004	0.003	0.009	0.007
BD +26 2606	-0.005	-0.007	0.001	0.012	0.010
BD +29 2091	0.003	-0.018	-0.005	0.006	0.014
BD +54 1216	0.007	0.000	0.003	0.019	0.017

Table 4. Statistical comparison.

	u'	g'	r'	i'	z'	All
RMS	0.007	0.008	0.005	0.014	0.015	0.010
Mean difference	0.003	–0.004	0.003	0.014	0.014	0.006

The statistics of the differences between synthetic and photometric magnitudes from Table 3 are presented by bandpass in Table 4. The root-mean-square (RMS) values range from 0.005 to 0.015 magnitude and the mean differences range from –0.004 to +0.014. The statistical results taken across all five Sloan bands are given in the last column of the Table. The overall RMS is only 0.010 magnitude and the overall mean difference is just +0.006.

The goals for the Sloan photometric standard star system (Smith et al. 2002) were given in percentages ranging from 1% to 1.5%. When converted to magnitudes these values are 0.016 for u', 0.011 for g', r', and i', and 0.016 for z'. Bohlin et al. (2014) quote an accuracy of 1% (0.011 magnitude) from the visible to the near-IR for the STIS/CALSPEC standard stars.

The combined uncertainty for STIS/CALSPEC and the Sloan photometric magnitudes considered in this study is the square root of the sum of the squares (RSS) of their separate uncertainties which follow: 0.019 for u', 0.016 for g', r', and i', and 0.019 for z'.

The RMS values in Table 4 are less than the RSS values in every band. Thus, the statistics of the observed differences between the magnitudes from the two sources are consistent with the combined uncertainties of those sources as stated therein.

The good agreement of the ultraviolet u' magnitudes (RMS, 0.007; mean, +0.003) is notable for three reasons. First, the Sloan observations were made from the ground where atmospheric extinction is very high at ultraviolet wavelengths. Second, the 1% accuracy quoted by Bohlin et al. applies to visible and near-IR wavelengths. Third, u' magnitudes are generally difficult to determine accurately as noted in several places, including Chonis and Gaskell (2008).

4. Comparison with Johnson-Cousins magnitudes

A study by Bohlin and Landolt (2015) is similar to that reported in this paper. However, they compared STIS/CALSPEC fluxes with magnitudes on the Johnson-Cousins photometric system. Based on data from 11 stars they found that the photometric observations and spectral fluxes agree to better than 0.010 magnitude for the B, V, R, and I bands. The central wavelengths of these bands range from 440 nm to 900 nm.

They also report that the white dwarf star BD +17 4708 varied by ~0.008 magnitude per year from 1986 through 1991. This star was included in the present study and Table 3 indicates a somewhat anomalous results in the i' and z' bands.

5. Conclusions and suggestions for future research with small telescopes

A comparison between synthetic magnitudes derived from STIS/CALSPEC fluxes and photometric magnitudes of Sloan standard stars was performed. The statistics of the differences are consistent with the stated or intended accuracies of both systems in all five Sloan bands.

The central wavelengths of those bands range from 355 to 893 nm. So, this study extends the comparison made by Bohlin and Landolt (2015) to the near-UV. It also confirms that the ground-based spectroscopy acquired by Oke (1990) was remarkably accurate.

One of the fundamental photometric standard stars was reported to be variable by Bohlin and Landolt. This star and the other white dwarfs used as standards are sufficiently bright that accurate photometry can be obtained with relatively small telescopes. Monitoring of the white dwarfs and other standard stars is suggested.

6. Acknowledgements

R. Bohlin (Space Telescope Science Institute) provided important information about the CALSPEC database. The comments of an anonymous reviewer were very helpful and improved the paper significantly.

References

Bohlin, R. C., Gordon, K. D., and Tremblay, P.-E. 2014, *Publ. Astron. Soc. Pacific*, **126**, 711 (DOI 10.1086/677655).

Bohlin, R. C., and Landolt, A. U. 2015, *Astron. J.*, **149**, 122 (DOI 10.1088/0004-6256/149/4/122).

Chonis, T. S., and Gaskell, C. M. 2008, *Astron. J.*, **135**, 264 (DOI 10.1088/0004-6256/135/1/264).

Fukugita, M., Ichikawa, T., Gunn, J. E., Doi, M., Shimasuka, K., and Schneider, D. P. 1996, *Astron. J.*, **111**, 1748 (DOI 10.1086/117915).

Oke, J. B. 1990, *Astron. J.*, **99**, 1621 (DOI 10.1086/115444).

Oke, J. B., and Gunn, J. E. 1983, *Astrophys. J.*, **266**, 713 (DOI 10.1086/160817).

Smith, J. A., *et al.* 2002, *Astron. J.*, **123**, 2121 (DOI 10.1086/339311).

Revised Light Elements of 78 Southern Eclipsing Binary Systems

Margaret Streamer
3 Lupin Place, Murrumbateman, NSW 2582, Australia; send email correspondence to m.streamatbigpond.com

Jeff Byron
18 Albuera Road, Epping, NSW 2121, Australia

David J. W. Moriarty
315 Main Road, Wellington Point, Qld 4160, Australia

Tom Richards
P.O. Box 323, Kangaroo Ground, Vic 3097, Australia

Bill Allen
83 Vintage Lane, RD3, Blenheim, New Zealand

Roy Axelsen
P.O. Box 706, Kenmore, Qld 4069, Australia

Col Bembrick
P.O. Box 1537, Bathurst, NSW 2795, Australia

Mark Blackford
25 Bambridge Street, Chester Hill, NSW 2162, Australia

Terry Bohlsen
Mirranook, Armidale, NSW 2350, Australia

David Herald
3 Lupin Place, Murrumbateman, NSW 2582, Australia

Roland Idaczyk
P.O. Box 22369, Khandallah, Wellington 6441, New Zealand

Stephen Kerr
22 Green Avenue, Glenlee, Qld 4711, Australia

Ranald McIntosh
139 Camerons Road, Marsden, Greymouth 7805, New Zealand

Yenal Ogmen
P.O. Box 756, Nicosia, North Cyprus via Mersin 10 Turkey

Jonathan Powles
40 Hensman Street, Latham ACT 2615, Australia

Peter Starr
841 Timor Road, Coonabarabran, NSW, 2357, Australia

George Stockham
77 Lindrum Crescent, Holt ACT 2615, Australia

Received October 14, 2014; revised December 2, 2014; accepted December 5, 2014

Abstract Since 2011, members of Variable Stars South have undertaken intensive time series observations and analysis of eclipsing binary systems, most of which are south of declination –40°. Many of them have not been observed in detail since their discovery 50 to 80 years ago. New or revised light elements are presented here for 60 systems and revised O–C values for a further 18 systems. A pulsating component has been discovered in four of the binary systems: RZ Mic, V632 Sco, V638 Sco, and LT Her.

1. Introduction

The Southern Eclipsing Binaries Programme of Variable Stars South (VSS) is a multi-purpose and ongoing campaign to observe and analyze bright eclipsing binary stars accessible to Southern Hemisphere observers. Despite their importance and ease of observation, many of them have not been observed in detail since their discovery, and others require follow-up work to check and extend existing studies (Richards 2013). When we began this study, many ephemerides were so far out of date that eclipses were not observed at the times predicted by the GCVS elements. It was necessary, therefore, to obtain accurate eclipse timings and update ephemeris elements. We present here new or revised light elements for 60 of the 150 targets selected for study and revised O–C values for a further 18 systems.

2. Observations and Analysis

The data reported here are based on observations from early in 2011 to the end of 2013. Time-series photometry was performed with the instruments given in Table 1. Each observer used NTP software such as DIMENSION 4 (Thinking Man Software 1992–2014) to synchronize their computer's clock to UTC. A fast cadence was used in acquiring the photometric data to ensure good coverage for accurate determination of the eclipse times of minima. Eclipses were observed over a period of several hours to cover both descent and ascent around the minimum.

All CCD imaging was done with Johnson V filters unless otherwise indicated. Bright targets were observed with DSLR cameras and the procedures described by Blackford and Schrader (2011) were used to convert magnitudes to the standard BVR_c system. All images were reduced using aperture photometry and the resulting magnitude data are all untransformed unless otherwise indicated. Times of minima were determined using the Kwee and van Woerden algorithm or the Polynomial fit in PERANSO (Vanmunster 2013). Where three or more times of primary minima were accurately determined for any target, a linear regression analysis was applied using the Linest function in MICROSOFT EXCEL to obtain improved light elements. Although more than three primary eclipse measurements are preferable for a regression analysis, the results are an improvement on previously published data. One exception to this was AS Mon, for which only secondary eclipses were used for the regression analysis.

To further refine the light elements, VSS data were combined with ASAS observations (Pojmański 1997) as follows.

The ASAS data were phase-folded using a period which, by eye, achieved a minimum corresponding to the VSS epoch calculated from the Linest function. For these phase-folded data an "ASAS pseudo time of minimum" and associated uncertainty were determined close to the median time for the ASAS data. A weighted regression analysis was then performed using the combined VSS and ASAS times of minima and associated uncertainties.

The magnitudes of the eclipses are given here as an observational aid, rather than as a definitive assessment of eclipse depths. Although determined from aperture photometry, the CCD data are untransformed and therefore subject to instrumental differences between observers. Nevertheless, for several targets our results give a more realistic indication of the depth of the eclipses compared to those determined from photographic plates or survey telescopes such as ASAS. Additionally, for some targets, the elimination of instrumental variations by using data sets from the same observer permitted discrimination between primary and secondary eclipses where their depths were very similar. This led to further investigation and reassessment of the period.

We also report the maximum magnitude of the uneclipsed portion of the light curve. Where we have no relevant data for this, we give an approximate average magnitude assessed from ASAS data. Again the latter are included solely as an observational aid.

For many of the targets we observed, the predictions for times of minima were quite inadequate, either because the period and/or epoch had not been updated since discovery or the prediction methods used by others were insufficiently robust. To improve predictions, one of us (Byron) developed a routine to find the period giving the minimum value to a "scatter" parameter of the phase-folded ASAS data. This revised period

Table 1. Equipment used by the different observers.

Observer	Initials This Paper	AAVSO Initials	Telescope	Camera
Streamer	MS	SFU	200 mm Meade Schmidt-Cassegrain	SBIG ST402 CCD
Streamer			350 mm Meade Schmidt-Cassegrain	SBIG ST8XME CCD
Allen	WA	AWH	410 mm Cassegrain f/15	SBIG STL 1001E
Axelsen	RA	ARX	230 mm Celestron C9.25 Schmidt-Cassegrain	Canon EOS 500D DSLR
Bohlsen	TB	BHQ	200 mm Vixen VC200L Cassegrain	SBIG ST10XME CCD
Herald	DH	—	400 mm Meade Schmidt-Cassegrain	SBIG STL-6303 CCD
Kerr	SK	KSH	80 mm William Optics Refractor	Meade DSI Pro II
Ogmen	YO	OYE	356 mm Meade LX200R (ACF)	SBIG ST8XME CCD
Moriarty	DM	MDJA	280 mm Celestron Schmidt-Cassegrain	SBIG ST8XME CCD
Moriarty			356 mm Celestron Edge HD 1400 aplanatic Schmidt-Cassegrain	Moravian G3-6303 CCD
Powles	JP	PJOC	254 mm Meade Schmidt-Cassegrain	Atik 383L+
Richards	TR	RIX	410 mm RCOS Ritchey-Chrétien	Apogee U9 CCD
Starr	PS	SPET	508 mm Planewave CDK	SBIG STL 6303

was used to determine a nominal epoch, corresponding to a calculated minimum, close to the mid-value of the HJD of the ASAS data. This combination of epoch and period provided an improved prediction of current times of minima compared to previously published elements.

3. Results and discussion

3.1. Revised epochs and periods

Epochs and periods have been revised for 25 eclipsing binaries for which we obtained sufficiently high quality primary eclipse data for regression analysis on the times of minima (Table 2). For most targets, a weighted regression analysis using the combined VSS and ASAS data resulted in little change in period or epoch, within the appropriate error limits, compared to the Linest determination using VSS data alone. These values are also recorded in Table 2 under VSS+ASAS (Wtd). It must be emphasized, however, that these values are only valid if there has been no change in period of the target during the years from when ASAS data and VSS data were taken. However, note that the Linest function tends to underestimate the uncertainties in the final results, whereas the weighted regression procedure is more conservative.

We determined new periods, or clarified discrepancies in the literature, for the following binary systems.

TZ Cru has a period of 2.091154 days, which is double that given in the *General Catalogue of Variable Stars* (GCVS; Kholopov *et al.* 1985) and used in the ASAS light curve. In the discovery paper, Bruna (1930) suggested that the period should probably be doubled. Our results clearly show primary eclipses at V magnitude 13.1 and secondary eclipses at V magnitude 13.0. There was no indication of a secondary eclipse occurring at or about phase 0.5 using the shorter period.

AS Mon. GCVS gives a period of 1.836486 days, which is that used by Alfonso-Garzon *et al.* (2012). Using these predictions we observed four eclipses. However, we now conclude that the period is double this and that we had observed three secondary (V magnitude 11.2) and one primary (V magnitude 11.3) eclipses. Our weighted regression analysis of only the secondary eclipses results in a period of 3.673106 days, close to that given by Diethelm (2012) and Pojmański with ASAS data.

V632 Sco. There is confusion in the literature concerning its period. Our period of 1.610156 days is similar to that reported by Malkov *et al.* (2006) and that used originally in the GCVS catalogue (1.610168 days). Dvorak (2004) reported a period of 3.2204 days. We confirmed the shorter period by identifying a shallow secondary minimum with V magnitude 11.2 which occurs at phase 0.5.

V5552 Sgr has a period of 1.347670 days, which is half that reported by Otero (2003) and that used for the ASAS light curve. A small secondary eclipse was also identified close to the time predicted with the shorter period. Although requiring confirmation, the latter may not occur at exactly 0.5 phase, thus suggesting apsidal motion in this system.

V536 Ara, GM Nor, and *CT Phe.* Even after a weighted regression analysis, our results for these three targets remained poorly aligned with ASAS data. The results determined from VSS data alone are given in Table 3. There are several possibilities to explain this mismatch of data. For example, there could be a small period change; a third body may be causing a small deviation in the period; there could be slight apsidal motion in the system. Further observations on these targets are required to confirm and clarify our results.

The individual times of minima for each target in Tables 2 and 3 are given in Table 4.

3.2. Improved epochs

For some targets there were insufficient numbers of observed minima for a linear regression analysis, but the epochs of the minima we observed permitted revisions that are an improvement on data previously available. These results are given in Table 5. The observed times of minima for some targets differed by several hours from predictions using the old epochs, some of which had not been updated since discovery. Multiple observations of these targets were sometimes needed before an eclipse was finally detected. For each target, a single current epoch is presented equal to the time of minimum derived from the best data set.

Even though we have limited data for the targets given in Table 5, some of them show interesting features and are highlighted below.

TZ CMa. We used the longer period as given by Kreiner (2004) although the latter has the primary and secondary eclipses incorrectly assigned. The eclipses are of almost equal depth (V magnitude, p = 10.6, s = 10.5). We found that the secondary eclipse occurs at phase 0.515 rather than 0.5, indicating an eccentric orbit for this system.

CZ Mic. Otero (2003) reported a period double that previously published (1.00944 days). We assessed ASAS data using both periods and believe the shorter period to be more appropriate and we give an updated epoch for the primary eclipse. We have recently identified the shallow secondary eclipse based on the shorter period.

EF Vel showed deep primary eclipses of V magnitude 15.4. Both the *International Variable Star Index* (VSX; Watson *et al.* 2014) and ASAS data plots give the primary eclipse depth as magnitude 13.7. We conclude the differences are due to the limitations of the equipment used by other observers.

3.3. O–C results from reliable predictions

For several targets, the published light elements provided reliable predictions for times of minima; these are listed in Table 6. We used the original GCVS periods and epochs to calculate O–C values unless otherwise indicated.

V339 Ara is worth noting. This binary system has shown virtually no change in period since the original GCVS light elements were published. However, the magnitudes reported for the system vary considerably. Our observations showed an out-of-eclipse V magnitude of 11.6, falling to 13.1 during primary eclipse. VSX gives the range from 10.2 to 10.8, as does the GCVS. ASAS data show the magnitude range from 11.0 to 11.7. These discrepancies are difficult to explain.

AR CMa, V526 Sgr, and *AO Vel* have eccentric orbits and show apsidal motion, and the times of minima for these are included in Table 6 as a resource for others studying these systems.

Table 2. Light elements revised by regression analysis of VSS eclipses using Linest and matched to ASAS data using a weighted (Wtd) regression analysis.

Target	#p	#s	Observers	Observation Span	Mag. Max.	Mag. p	Mag. s	Revised Period (days) VSS data (Linest)	Revised Period (days) VSS + ASAS (Wtd)	Revised Epoch (HJD) VSS data (Linest)	Revised Epoch (HJD) VSS + ASAS (Wtd)
XY Ant	3	1	MS	2012-03-2013-05	9.0*	10.7	10.6	2.1803412 ±0.0000014	2.1803419 ±0.0000005	2456438.9951 ±0.0002	2456438.9952 ±0.0005
V881 Ara	3	1	MS	2012-05-2013-07	10.1	10.7	10.6	2.4188846 ±0.0000014	2.4188894 ±0.0000005	2456064.1238 ±0.0001	2456064.1237 ±0.0002
DI Cen	3	—	MS, PS	2012-04-2012-05	11.6*	12.4	—	3.549563 ±0.000057	3.5495684 ±0.0000015	2456045.9981 ±0.0003	2456045.9980 ±0.0004
V775 Cen	4	3	DM	2011-08-2013-06	9.7	10.3	9.9	0.6636414 ±0.0000009	0.6636413 ±0.0000001	2455808.8962 ±0.0005	2455808.8963 ±0.0005
V777 Cen	3	—	MS, TR	2012-05-2013-02	10.9	11.8	—	1.7759917 ±0.0000053	1.7759942 ±0.0000006	2456313.0666 ±0.0005	2456313.0666 ±0.0005
SZ Cru	5	2	DM, TR	2011-06-2013-05	11.2	12.3	11.5	1.9743162 ±0.0000015	1.9743156 ±0.0000006	2456430.0678 ±0.0003	2456430.0678 ±0.0005
TZ Cru	8	5	DM, MS, TR	2012-03-2013-06	12.4	13.1	13.0	2.0911644 ±0.0000046	2.0911536 ±0.0000008	2456057.9998 ±0.0005	2456058.0008 ±0.0005
AA Cru	4	2	DM, MS	2011-06-2013-04	11.0	11.5	11.4	3.7876324 ±0.0000036	3.7876337 ±0.0000007	2456070.0280 ±0.0003	2456070.0279 ±0.0009
BE Cru	4	3	DM, TR	2011-06-2013-05	12.7	13.5	12.7	2.2210130 ±0.0000032	2.2210076 ±0.0000017	2456084.0376 ±0.0002	2456084.0378 ±0.0004
RU Gru	4	1	DM, MS	2011-08-2013-10	11.1	11.8	11.1	1.8932001 ±0.0000018	1.8931963 ±0.0000009	2455805.1032 ±0.0005	2455805.1040 ±0.0007
TT Hor	6	2	DM, MS, TR	2011-11-2013-12	10.9	11.6	11.0	2.6082143 ±0.0000033	2.6082123 ±0.0000017	2456267.0788 ±0.0003	2456267.0789 ±0.0014
KZ Lib	3	—	MS	2012-05-2012-07	11.2	13.0	—	1.2387368 ±0.0000001	1.2387368 ±0.0000008	2456086.1355 ±0.0001	2456086.1355 ±0.0003
RZ Mic	3	1	MS	2011-10-2013-09	11.3*	12.3	11.6	3.9830610 ±0.0000048	3.9830361 ±0.0000029	2456543.9738 ±0.0005	2456543.9738 ±0.0005
CY Mic	3	—	MS, PS	2012-05-2012-08	11.7	12.4	—	1.6287250 ±0.0000037	1.6287358 ±0.0000007	2456153.0395 ±0.0001	2456153.0397 ±0.0005
AS Mon	1	3	MS	2011-12-2013-01	10.6	11.3	11.2	3.6731149 ±0.0000012	3.6731059 ±0.0000017	2456308.1235 ±0.0001	2456308.1227 ±0.0006
HM Pup	3	—	DM, MS	2012-03-2013-12	10.8	14.3	10.9	2.5897029 ±0.0000029	2.5897006 ±0.0000018	2456306.9982 ±0.0003	2456306.9982 ±0.0004
V849 Sgr	4	—	MS, TR	2011-09-2013-08	12.6	13.7	—	2.9506535 ±0.0000001	2.9506519 ±0.0000019	2455818.9621 ±0.0001	2455818.9623 ±0.0014
V5552 Sgr	4	1	DM, MS	2012-07-2013-06	12.9	13.6	13.0	1.3476614 ±0.0000060	1.3476697 ±0.0000014	2456152.0751 ±0.0007	2456152.0752 ±0.0010
V490 Sco	3	3	DM, MS	2012-07-2013-08	11.9	12.3	12.3	3.003763 ±0.000054	3.0037555 ±0.0000023	2456128.9342 ±0.0003	2456128.9341 ±0.0005
V626 Sco	5	—	DM, MS, TR	2012-06-2012-08	11.3	12.1	11.6	1.0336825 ±0.0000010	1.0336847 ±0.0000004	2456510.9553 ±0.0002	2456510.9555 ±0.0007
V632 Sco	4	—	TB, DM, MS	2012-07-2013-09	11.1	11.9	11.2	1.610148 ±0.000015	1.6101555 ±0.0000011	2456476.8968 ±0.0004	2456476.8965 ±0.0011
V634 Sco	4	1	DM, MS	2011-07-2013-06	11.6*	12.1	12.0	1.2240290 ±0.0000010	1.2240290 ±0.0000007	2455769.9765 ±0.0001	2455769.9763 ±0.0003
V638 Sco	3	—	DM, MS	2011-08-2013-08	10.8	11.7	—	2.3582339 ±0.0000085	2.3582243 ±0.0000014	2456114.1140 ±0.0011	2456114.1153 ±0.0013
LU Tel	3	1	DM	2012-06-2013-09	12.4	14.0	12.5	1.5717378 ±0.0000020	1.5717431 ±0.0000010	2456096.1639 ±0.0004	2456096.1637 ±0.0006
AW Vel	6	2	DM, MS, TR	2012-05-2013-03	10.7	12.2	10.8	1.992458 ±0.000038	1.9924771 ±0.0000005	2456274.1503 ±0.0003	2456274.1502 ±0.0003

* ASAS data

Table 3. Light elements revised by regression analysis of VSS eclipses using Linest with a poor match to ASAS data.

Target	#p	#s	Observers	Observation Span	Mag. Max.	Mag. p	Mag. s	Revised Period (days)	Revised Epoch (HJD)
V536 Ara	3	—	MS	2012-05–2013-07	11.7*	13.2	—	2.3740622 ±0.0000016	2456062.1003 ±0.0002
GM Nor	3	—	TB, DM, DH	2011-07–2013-04	10.6	11.1	—	1.8846063 ±0.0000017	2456112.0091 ±0.0002
CT Phe	3	—	WA, TB, MS, TR	2011-09–2013-11	11.3	11.8	—	1.2608260 ±0.0000025	2456195.0750 ±0.0005

* ASAS data

Table 4. Times of minima observed for targets listed in Tables 2 and 3.

Target	Eclipse Type*	Observers	Time of Minimum (HJD)	Error (day)	Target	Eclipse Type*	Observers	Time of Minimum (HJD)	Error (day)
XY Ant	p	MS	2456016.0090	0.0001	KZ Lib	p	MS	2456074.9868	0.0006
	p	MS	2456087.9600	0.0001		p	MS	2456086.1355	0.0005
	p	MS	2456438.9951	0.0001		p	MS	2456131.9687	0.0005
	s	MS	2456437.9050	0.0001	RZ Mic	p	MS	2455846.9381	0.0050
V536 Ara	p	MS	2456062.1004	0.0001		s	MS	2456537.9960	0.0100
	p	MS	2456138.0701	0.0001		p	MS	2456539.9903	0.0002
	p	MS	2456487.0574	0.0001		p	MS	2456543.9743	0.0001
V881 Ara	p	MS	2456064.1238	0.0004	CY Mic	p	PS	2456055.3161	0.0001
	p	MS	2456139.1093	0.0003		p	MS	2456153.0397	0.0008
	p	MS	2456172.9736	0.0003		p	MS	2456170.9554	0.0001
	s	MS	2456478.9625	0.0001	AS Mon	s	MS	2455904.0808	0.0001
DI Cen	p	MS	2456045.9983	0.0001		p	MS	2455949.9946	0.0001
	p	PS	2456053.0969	0.0001		s	MS	2455995.9088	0.0001
	p	MS	2456077.9442	0.0001		s	MS	2456308.1235	0.0014
V775 Cen	p	DM	2455798.9418	0.0019	GM Nor	p	DM	2456112.0093	0.0001
	p	DM	2455808.8956	0.0017		p (UF)	DH	2456128.9704	0.0004
	p	DM	2456075.0171	0.0014		p	TB	2456406.0077	0.0003
	s	DM	2456458.9343	0.0030	CT Phe	p	MS	2456195.0745	0.0003
	p	DM	2456459.9282	0.0009		p	MS	2456205.1621	0.0002
V777 Cen	p	MS	2456051.9958	0.0010		p	TR	2456601.0608	0.0015
	p	MS	2456313.0671	0.0007	HM Pup	p	MS	2455991.0542	0.0005
	p	TR	2456329.0501	0.0001		p	MS	2456306.9986	0.0005
SZ Cru	p	DM	2455748.9281	0.0026		p	MS	2456628.1211	0.0002
	p	DM	2455750.9035	0.0001	V849 Sgr	p	MS	2455818.9621	0.0025
	p (R_c)	TR	2456355.0435	0.0004		p	MS	2456116.9781	0.0026
	s (R_c)	TR	2456356.0203	0.0070		p (R_c)	TR	2456465.1552	0.0036
	p (R_c)	TR	2456422.1709	0.0001		p	MS	2456530.0696	0.0028
	p	DM	2456430.0675	0.0017	V5552 Sgr	p	MS	2456152.0752	0.0016
	s	DM	2456431.0349	0.0014		p	MS	2456174.9864	0.0014
TZ Cru	p	DM	2455988.9895	0.0030		p	MS	2456201.9374	0.0028
	p	MS	2456055.9094	0.0012		s	MS	2456203.9666	0.0057
	s	MS	2456056.9546	0.0015		p	DM	2456466.0804	0.0032
	p	DM	2456058.0005	0.0001	V490 Sco	p	DM	2456128.9343	0.0001
	s	DM	2456059.0453	0.0021		p	MS	2456140.9489	0.0001
	p	MS	2456078.9121	0.0012		p	MS	2456158.9719	0.0015
	s	DM	2456100.8696	0.0002		s	DM	2456508.9095	0.0001
	s	DM	2456123.8726	0.0017		s	DM	2456520.9256	0.0002
	p (R_c)	TR	2456354.9457	0.0016		s	DM	2456535.9399	0.0001
	s (R_c)	TR	2456355.9899	0.0019	V626 Sco	p	MS	2456115.0549	0.0011
	p (R_c)	TR	2456357.0371	0.0016		p	TR	2456484.0796	0.0013
	p	MS	2456377.9478	0.0017		p	DM	2456509.9212	0.0014
	p	MS	2456446.9553	0.0014		p	DM	2456510.9554	0.0013
	s	MS	2456448.0006	0.0016		p	DM	2456513.0230	0.0016
AA Cru	p	DM	2455732.9282	0.0002	V632 Sco	p	MS	2456476.8964	0.0020
	p	DM	2455767.0177	0.0004		p	DM	2456484.9479	0.0026
	s	MS	2455996.1702	0.0001		p	DM	2456542.9124	0.0017
	p	DM	2456070.0282	0.0001		p	TB	2456542.9133	0.0020
	s	DM	2456124.9488	0.0001	V634 Sco	p	DM	2455769.9766	0.0008
	p	DM	2456410.9147	0.0001		p	MS	2456124.9447	0.0002
BE Cru	p	DM	2456084.0372	0.0001		p	MS	2456146.9772	0.0002
	p	DM	2456092.9221	0.0004		p	MS	2456452.9845	0.0004
	p	DM	2456112.9108	0.0001		s	MS	2456460.9402	0.0003
	s	DM	2456384.9907	0.0022	V638 Sco	p	DM	2456114.1139	0.0021
	s	DM	2456404.9534	0.0126		p	MS	2456481.9993	0.0018
	p	DM	2456406.0845	0.0003		p	MS	2456515.0130	0.0033
	s	DM	2456416.1055	0.0004	LU Tel	p	DM	2455773.9577	0.0012
RU Gru	p	MS	2455805.1025	0.0012		p	DM	2456096.1644	0.0009
	p	MS	2455858.1135	0.0016		s	DM	2456537.0117	0.0032
	p	DM	2456533.9854	0.0013		p	DM	2456540.9654	0.0011
	p	DM	2456586.9947	0.0015	AW Vel	p	MS	2456274.1502	0.0005
TT Hor	p	MS	2456267.0793	0.0031		p	MS	2456282.1202	0.0006
	p	MS	2456280.1201	0.0027		p	DM	2456292.0821	0.0009
	p	TR	2456280.1193	0.0025		p	TR	2456292.0827	0.0007
	p	MS	2456567.0232	0.0035		p	MS	2456296.0669	0.0006
	p	MS	2456580.0651	0.0035		p	TR	2456296.0677	0.0006
	p	MS	2456627.0122	0.0049					

*(R_c) indicates minima observed with Cousins R filter; (UF) are unfiltered observations.

Table 5. Current epochs of targets with limited eclipse data.

Target	Eclipse Type**	Observers	Magnitude Uneclipsed	Mag. p	Mag. s	Ref. Period (days)	Current Epoch (HJD)	Error (day)
TZ CMa	p	MS, TR	10.1	10.6	—	3.822884	2456299.0585	0.0013
TZ CMa	s	MS, TR	10.1	—	10.5	3.822884	2456299.1085	0.0013
DO Car	p	DM, TR	9.0	9.3	—	3.85194	2456066.9614	0.0003
GP Car	s	MS	11.4*	—	11.9	2.464172	2456327.0850	0.0003
V594 Car	p	TB, DM	10.1	10.8	—	3.165830	2456402.8984	0.0025
ST Cen	s	DM	10.5	—	11.2	1.22339	2456435.9265	0.0011
V471 CrA	p	MS, TR	12.8	13.2	—	2.486044	2456532.9969	0.0028
AF Cru	p	DM, MS, TR	9.8	10.6	9.9	1.895685	2456099.0083	0.0002
W Gru	p	SK, MS	8.9*	9.5	—	2.968535	2456181.0712	0.0001
LT Her	p (UF)	YO	10.6	11.0	—	1.084033	2455752.3186	0.0004
GK Lib	p	PS, TR	12.4*	14.5	12.7	2.116458	2456058.0799	0.0001
FV Lup	p	DM, MS	10.2*	11.2	—	4.103084	2456462.0946	0.0001
RR Men	p	DM	11.4*	12.9	—	2.599540	2456290.9499	0.0002
CZ Mic	p	MS, TR	12.7	13.3	—	1.00944	2456470.0727	0.0001
VX Nor	p	DM, MS	11.4	12.0	—	2.082440	2456129.1177	0.0002
V384 Nor	p	DM	10.1*	10.7	—	3.97413	2456109.0133	0.0003
RV Pic	p	DM	9.6*	11.9	—	3.97178	2455887.0659	0.0003
AH Pup	p	MS, TR	11.6*	12.6	—	2.02485	2456250.1528	0.0005
KX Pup	p	MS	12.3	12.7	—	2.146777	2456302.0876	0.0025
SV Pyx	p	MS, TR	10.7*	11.4	—	1.446429	2456061.8969	0.0005
V789 Sgr	p	MS, TR	12.4*	12.7	—	2.55234	2456463.1385	0.0005
V2351 Sgr	p	MS	10.1*	10.7	—	3.748819	2456143.07195	0.0030
V606 Sco	p	MS	11.7	12.5	—	2.6857	2456165.9949	0.0001
V1226 Sco	p	MS	10.6	10.8	—	2.08355	2456458.9573	0.0007
V1270 Sco	p	MS	9.2*	9.7	—	4.243190	2456179.937	0.001
V356 Sct	p	MS	11.8*	12.4	—	2.1229	2455798.9312	0.0001
EQ TrA	p	SK, DM, MS	8.4*	9.1	—	2.709149	2456067.9472	0.0001
RV Vel	p	DM	9.9*	10.5	—	4.82105	2456060.9268	0.0004
AR Vel	p	MS	12.3*	12.9	—	3.212764	2456063.9098	0.0024
EF Vel	p	MS	13.2*	15.4	—	3.0696	2456077.0065	0.0022
EL Vel	p (R_c)	TR	11.3*	12.4	—	2.758338	2456340.0922	0.0002
ET Vel	p	TR	11.1*	11.9	—	3.080858	2456323.0803	0.0003
FV Vel	p	TB, MS	10.9	12.0	—	1.521131	2455962.1014	0.0001

* ASAS data. **(R_c) indicates minima observed with Cousins R filter. (UF) are unfiltered observations.

Zasche (2012) revised the period for AR CMa to double that previously reported and showed that the system had an apsidal motion of a period of 44 years.

Oosterhoff and van Houten (1949) reported that AO Vel has an orbital eccentricity of 0.12 with the line of apsides moving with a period of about 50 years. It has since been reported as a quadruple system formed by two double-lined spectroscopic binaries (Gonzalez et al. 2006). Our data, obtained over a six-month time span, showed the secondary eclipse at phase 0.698 using GCVS elements.

V526 Sgr has an orbital eccentricity of 0.22 and apsidal period of about 156 years. We used the original GCVS elements for O–C values.

4. New pulsating systems

We announce the discovery of four new systems that exhibit δ Scuti-type pulsations in their light curves. These are RZ Mic, V632 Sco, V638 Sco, and LT Her. The light curves of Ebbighausen and Penegor (1974) clearly showed the presence of pulsations for LT Her but they did not comment on their nature.

Of the 78 eclipsing binary systems that we report in this paper, we have now found seven that contain a pulsating component. Our preliminary characterization of AW Vel, HM Pup, and TT Hor has already been published (Moriarty et al. 2013). We shall be presenting our data on the four new systems in another paper.

5. Acknowledgements

David Moriarty acknowledges the support of a grant for the purchase of a telescope from the Edward Corbould Research Fund of the Astronomical Association of Queensland. Margaret Streamer and David Moriarty acknowledge grants from Variable Stars South to purchase software. Margaret Streamer acknowledges the American Association of Variable Star Observers (AAVSO) for the loan of the SBIG ST402 during 2011. This research has made use of the International Variable Star Index (VSX) database, operated at AAVSO, Cambridge, Massachusetts, USA.

References

Alfonso-Garzon, J., Domingo, A., Mas-Hesse, J. M., and Gimenez, A. 2012, *Astron. Astrophys.*, **548A**, 79.
Blackford, M. G., and Schrader, G. 2011, *Variable Stars South Newsletter*, **2**, 8.
Bruna, P. P. 1930, *Bull. Astron. Inst. Netherlands*, **6**, 45.
Chen, K-Y. 1975, *Acta Astron.*, **25**, 89.

Table 6. Times of minima and O–C values from observed eclipses.

Target	Eclipse Type*	Observers	Cycle	Time of Minimum (HJD)	Error	O–C (day)	Ref. Period (days)	Ref. Epoch (HJD)	Eclipse Mag.
V339 Ara	p	MS	11242	2456115.9705	0.0013	0.0001	2.438853	2428698.385	13.1
	p	MS	11267	2456176.9411	0.0015	–0.0007			13.1
TU CMa	p	MS	25700	2455961.9999	0.0001	–0.0104	1.1278041	2426977.445	10.4
	s	TR	26017.5	2456320.0804	0.0007	–0.0078	—	—	10.0
AR CMa	p	MS	1391[a]	2455893.0072	0.001	0.0106	2.3322242	2452648.8727	11.6
	p	MS	1436[a]	2455997.9547	0.002	0.0081	—	—	11.6
DQ Car	s	MS	1882.5[a]	2456088.9324	0.0009	–0.0001	1.733678	2452825.284	11.6
	p	MS	1841[a]	2456016.9853	0.0006	0.0001	—	—	11.7
	s	MS	2009.5[a]	2456309.1068	0.0008	–0.0031	—	—	11.6
	s	MS	2047.5[a]	2456374.9863	0.0008	–0.0034	—	—	11.6
	p	MS	2085[a]	2456439.9985	0.0007	–0.0042	—	—	11.7
V762 Cen	p	MS	1076	2456069.0078	0.0001	–0.0042	3.367895	2452445.157	12.4
	p	DM	1084	2456095.9517	0.0001	–0.0035	—	—	12.4
CW Eri	p	SK	5329[b]	2455807.1603	0.0002	–0.0027	2.728371	2441267.6756	8.9
	s	MS	5478.5[b]	2456215.0591	0.0004	0.0046	—	—	8.7
VZ Hya	p	MS	402	2456033.9261	0.0001	–0.0006	2.904301	2454866.3976	9.5
FZ Lup	p	SK, MS	791[c]	2456086.9599	0.0001	–0.0085	4.534625	2452500.08	10.7
AN Mon	p	MS	1433[c]	2456006.9913	0.0003	–0.0031	2.4458	2452502.163	12.9
	s	MS	1576.5[c]	2456357.9660	0.0005	–0.0007	—	—	12.1
AQ Mon	p	JP, MS	1618[c]	2456620.1341	0.0001	0.0010	2.545551	2452501.431	11.3
V648 Ori	p	TR	5068[c]	2456622.1442	0.0011	–0.0002	0.8132364	2452500.6623	12.4
	p	TR	5074[c]	2456627.0225	0.0007	–0.0013	—	—	12.4
	p	JP	5090[c]	2456640.0351	0.0003	–0.0005	—	—	12.4
LT Pup	p	MS	18030	2455979.9606	0.0001	–0.0069	1.642681	2426362.429	13.1
V526 Sgr	p	MS	17769	2456160.0188	0.0012	–0.0826	1.919411	2422054.0856	10.3
	s	MS	17782.5	2456186.1006	0.0014	0.0872	—	—	10.0
V457 Sco	p	DM, MS	14420	2456129.0399	0.0001	–0.0097	2.00738	2427182.63	11.3
V569 Sco	p	MS	2983[a]	2455797.0394	0.0001	0.0020	1.04724351	2452673.11	11.5
	s	MS	3350.5[a]	2456181.9007	0.0003	0.0013	—	—	11.4
	p	MS	3301[a]	2456130.0628	0.0003	0.0020	—	—	11.5
	p (Tfd)	RA	3659[a]	2456504.9766	0.0013	0.0026	—	—	11.5
	p (Tfd)	RA	3660[a]	2456506.0237	0.0014	0.0024	—	—	11.5
	p	MS	3660[a]	2456506.0233	0.0006	0.0020	—	—	11.5
CE Scl	p	SK, MS	3174	2455865.0515	0.0001	0.0070	2.277687	2448635.666	9.9
	p	SK, MS	3318	2456193.0372	0.0001	0.0057	—	—	9.9
AO Vel	p	MS	19946	2455876.0319	0.0002	0.2812	1.5845993	2424269.333	9.8
	p	MS	20189	2456261.0868	0.0001	0.2786	—	—	9.8
	s (R_c)	TR	20239.5	2456341.1439	0.0002	0.3134	—	—	9.8
	s	MS	20261.5	2456376.0067	0.0001	0.3150	—	—	9.7
	s	MS	20273.5	2456395.0206	0.0001	0.3137	—	—	9.7
BC Vel	p	MS	2991[c]	2456011.0128	0.0013	–0.0068	1.173598	2452500.788	11.6
	p	MS	3216[c]	2456275.0716	0.0021	–0.0076	—	—	11.6
	p (R_c)	TR	3268[c]	2456336.0982	0.0018	–0.0081	—	—	11.7
	p	MS	3308[c]	2456383.0434	0.0014	0.0067	—	—	11.6

Notes: a, Zasche (2012) light elements; b, Chen (1975) light elements; c, Kreiner (2004) light elements.
*(R_c) indicates minima observed with Cousins R filter; (Tfd) indicates transformed data obtained with DSLR camera.

Diethelm, R. 2012, *Inf. Bull. Var. Stars*, No. 6029, 1.
Dvorak, S.W. 2004, *Inf. Bull. Var. Stars*, No. 5542, 1.
Ebbighausen, E. G., and Penegor, G. 1974, *Publ. Astron. Soc. Pacific*, **86**, 203.
Gonzalez, J. F., Hubrig, S., Nesvacil, N., and North, P. 2006, *Astron. Astrophys.*, **449**, 327.
Kholopov, P. N., et al. 1985, *General Catalogue of Variable Stars*, 4th Ed., Moscow.
Kreiner, J. M. 2004, *Acta Astron.*, **54**, 207.
Malkov, O. Y., Oblak, E., Snegireva, E. A., and Torra, J. 2006, *Astron. Astrophys.*, **446**, 785.
Moriarty, D. J. W., Bohlsen, T., Heathcote, B., Richards, T., and Streamer, M. 2013, *J. Amer. Assoc. Var. Star Obs*, **41**, 182.
Oosterhoff, P. Th., and van Houten, C. J. 1949, *Bull. Astron. Inst. Netherlands*, **11**, 63.
Otero, S. A. 2003, *Inf. Bull. Var. Stars*, No. 5480, 1.
Pojmański, G. 1997, *Acta Astron.*, **47**, 467.
Richards, T. 2013, Southern eclipsing binaries programme of the Variable Stars South group (http://www.variablestarssouth.org/research/variable-types/eclipsing-binaries).
Thinking Man Software. 1992–2014, DIMENSION 4 software (http://www.thinkman.com/dimension4/).
Vanmunster, T. 2013, Light Curve and Period Analysis Software, PERANSO v.2.50 (http://www.peranso.com/).
Watson, C., Henden, A. A., and Price, C. A. 2014, AAVSO International Variable Star Index VSX (Watson+, 2006–2014; http://www.aavso.org/vsx).
Zasche, P. 2012, *Acta Astron.*, **62**, 97.

Recent Maxima of 67 Short Period Pulsating Stars

Gerard Samolyk
P.O. Box 20677, Greenfield, WI 53220; gsamolyk@wi.rr.com

Received January 8, 2015; accepted January 8, 2015

Abstract This paper contains times of maxima for 67 short period pulsating stars (primarily RR Lyrae and δ Scuti stars). These maxima represent the CCD observations received by the AAVSO Short Period Pulsator (SPP) section in 2014.

1. Recent observations

The accompanying list contains times of maxima calculated from CCD observations made by participants in the AAVSO's Short Period Pulsator (SPP) Section. This list will be web-archived and made available through the AAVSO ftp site at ftp:ftp.aavso.org/public/datasets/gsamoj431.txt. The error estimate is included. RR Lyr stars in this list, along with data from earlier AAVSO publications, are included in the GEOS database at: http://rr-lyr.ast.obs-mip.fr/dbrr/dbrr-V1.0_0.php. This database does not include δ Scuti stars. These observations were reduced by the writer using the PERANSO program (Vanmunster 2007). Column F indicates the filter used.

The linear elements in the *General Catalogue of Variable Stars* (GCVS; Kholopov *et al.* 1985) were used to compute the O–C values for most stars. For a few exceptions where the GCVS elements are missing or are in significant error, light elements from another source are used: RZ Cap and DG Hya (Samolyk 2010), and VY LMi (Henden and Vidal-Sainz 1997).

References

Henden, A. A., and Vidal-Sainz, J. 1997, *Inf. Bull. Var. Stars.*, No. 4535, 1.

Kholopov, P. N., *et al.* 1985, *General Catalogue of Variable Stars*, 4th ed., Moscow.

Samolyk, G., 2010, *J. Amer. Assoc. Var. Stars*, **38**, 12.

Vanmunster, T. 2007, PERANSO period analysis software (http://www.peranso.com).

Table 1. Recent times of maxima of stars in the AAVSO Short Period Pulsator program.

Star	JD (max) Hel. 2400000+	Cycle	O–C	F	Observer	Error	Star	JD (max) Hel. 2400000+	Cycle	O–C	F	Observer	Error
SW And	56667.2931	87128	–0.4226	V	T. Arranz	0.0007	TZ Aur	56687.3511	93917	0.0139	V	T. Arranz	0.0006
SW And	56674.3675	87144	–0.4247	V	T. Arranz	0.0006	TZ Aur	56699.4895	93948	0.0104	V	T. Arranz	0.0005
SW And	56683.6561	87165	–0.4240	V	R. Sabo	0.0019	TZ Aur	56710.4588	93976	0.0128	V	T. Arranz	0.0006
SW And	56705.3260	87214	–0.4258	V	T. Arranz	0.0008	TZ Aur	56723.3848	94009	0.0135	V	T. Arranz	0.0006
SW And	56923.8071	87708	–0.4307	V	G. Samolyk	0.0012	TZ Aur	56923.9222	94521	0.0135	V	G. Samolyk	0.0007
SW And	56960.5106	87791	–0.4364	V	T. Arranz	0.0006	BH Aur	56668.3176	30513	0.0075	V	T. Arranz	0.0011
SW And	56988.3755	87854	–0.4351	V	T. Arranz	0.0009	BH Aur	56674.7013	30527	0.0060	V	R. Sabo	0.0012
SW And	57000.3176	87881	–0.4346	V	T. Arranz	0.0012	BH Aur	56677.4400	30533	0.0081	V	T. Arranz	0.0009
SW And	57019.3288	87924	–0.4414	V	T. Arranz	0.0007	BH Aur	56680.6275	30540	0.0030	V	K. Menzies	0.0009
XX And	56894.7450	24638	0.2638	V	K. Menzies	0.0012	BH Aur	56694.3128	30570	0.0056	V	T. Arranz	0.0008
XX And	56925.8233	24681	0.2639	V	R. Sabo	0.0017	BH Aur	56699.3297	30581	0.0055	V	T. Arranz	0.0007
XX And	56951.8430	24717	0.2647	V	R. Sabo	0.0016	BH Aur	56708.4515	30601	0.0055	V	T. Arranz	0.0007
ZZ And	56679.6461	57438	0.0236	V	R. Sabo	0.0021	BH Aur	56952.9142	31137	0.0041	V	G. Samolyk	0.0013
ZZ And	56999.6125	58015	0.0247	V	R. Sabo	0.0017	BH Aur	56999.8910	31240	0.0036	V	R. Sabo	0.0020
AC And	56961.4013	46644	0.0152	V	T. Arranz	0.0021	BH Aur	57020.4154	31285	0.0040	V	T. Arranz	0.0007
AC And	56966.4233	46654	–0.2140	V	T. Arranz	0.0009	RS Boo	56742.9222	39679	–0.0004	V	G. Samolyk	0.0011
AC And	56986.3651	46692	–0.2270	V	T. Arranz	0.0016	RS Boo	56779.9093	39777	0.0075	V	R. Sabo	0.0011
AT And	56924.8156	23636	–0.0019	V	G. Samolyk	0.0024	RS Boo	56793.4871	39813	0.0011	V	T. Arranz	0.0005
AT And	56949.4889	23676	–0.0052	V	T. Arranz	0.0025	RS Boo	56801.4094	39834	–0.0007	V	T. Arranz	0.0007
AT And	56962.4432	23697	–0.0061	V	T. Arranz	0.0014	RS Boo	56830.4705	39911	0.0053	V	T. Arranz	0.0006
AT And	56980.3316	23726	–0.0083	V	T. Arranz	0.0009	RS Boo	56847.4502	39956	0.0047	V	T. Arranz	0.0005
AT And	56999.4480	23757	–0.0162	V	T. Arranz	0.0015	ST Boo	56746.7831	60366	0.0975	V	G. Samolyk	0.0016
SW Aqr	56901.3660	69244	–0.0005	V	T. Arranz	0.0007	ST Boo	56807.7759	60464	0.1058	V	R. Sabo	0.0021
TZ Aqr	56934.6749	33939	0.0116	V	R. Sabo	0.0018	ST Boo	56826.4514	60494	0.1126	V	T. Arranz	0.0008
YZ Aqr	56915.7857	39047	0.0718	V	R. Sabo	0.0017	ST Boo	56831.4292	60502	0.1121	V	T. Arranz	0.0006
AA Aqr	56956.6942	59466	–0.1495	V	G. Samolyk	0.0013	ST Boo	56849.4762	60531	0.1126	V	T. Arranz	0.0007
BR Aqr	56890.9242	39958	–0.1959	V	G. Samolyk	0.0014	ST Boo	56854.4535	60539	0.1116	V	T. Arranz	0.0008
CY Aqr	56930.4705	370620	0.0140	V	T. Arranz	0.0002	ST Boo	56877.4803	60576	0.1136	V	T. Arranz	0.0007
TZ Aur	56680.3007	93899	0.0136	V	T. Arranz	0.0005	SW Boo	56699.8730	27513	0.4144	V	K. Menzies	0.0009
TZ Aur	56681.4765	93902	0.0144	V	T. Arranz	0.0007	SW Boo	56790.7727	27690	0.4196	V	R. Sabo	0.0013
TZ Aur	56682.6518	93905	0.0147	V	R. Sabo	0.0008	SZ Boo	56773.8500	55724	0.0085	V	G. Samolyk	0.0012

Table continued on following pages

Table 1. Recent times of maxima of stars in the AAVSO Short Period Pulsator program, cont.

Star	JD (max) Hel. 2400000+	Cycle	O–C	F	Observer	Error	Star	JD (max) Hel. 2400000+	Cycle	O–C	F	Observer	Error
SZ Boo	56808.8808	55791	0.0103	V	R. Sabo	0.0019	XZ Cyg	56913.3727	27408	–2.3809	V	T. Arranz	0.0013
SZ Boo	56811.4941	55796	0.0095	V	T. Arranz	0.0011	XZ Cyg	56920.3720	27423	–2.3821	V	T. Arranz	0.0004
SZ Boo	56839.7256	55850	0.0088	V	R. Sabo	0.0015	XZ Cyg	56926.4344	27436	–2.3868	V	T. Arranz	0.0006
TV Boo	56753.8398	102842	0.0951	V	G. Samolyk	0.0020	XZ Cyg	56927.3685	27438	–2.3861	V	T. Arranz	0.0006
TV Boo	56790.4232	102959	0.1091	V	T. Arranz	0.0011	XZ Cyg	56934.3681	27453	–2.3870	V	T. Arranz	0.0006
TV Boo	56795.3981	102975	0.0830	V	T. Arranz	0.0008	XZ Cyg	56948.3537	27483	–2.4024	V	T. Arranz	0.0006
TV Boo	56799.4935	102988	0.1151	V	T. Arranz	0.0007	XZ Cyg	56949.2880	27485	–2.4015	V	T. Arranz	0.0007
TV Boo	56820.4342	103055	0.1144	V	T. Arranz	0.0007	XZ Cyg	56955.3513	27498	–2.4053	V	T. Arranz	0.0006
TV Boo	56850.4303	103151	0.1048	V	T. Arranz	0.0018	XZ Cyg	56962.3584	27513	–2.3987	V	T. Arranz	0.0008
TV Boo	56857.6291	103174	0.1147	V	K. Menzies	0.0021	DM Cyg	56862.7597	34012	0.0754	V	R. Sabo	0.0011
TW Boo	56696.8891	55997	–0.0785	V	K. Menzies	0.0007	DM Cyg	56877.8795	34048	0.0802	V	R. Sabo	0.0015
TW Boo	56737.8733	56074	–0.0793	V	G. Samolyk	0.0011	DM Cyg	56890.4748	34078	0.0797	V	T. Arranz	0.0007
TW Boo	56796.4220	56184	–0.0807	V	T. Arranz	0.0009	DM Cyg	56893.4113	34085	0.0772	V	T. Arranz	0.0006
TW Boo	56812.3891	56214	–0.0818	V	T. Arranz	0.0005	DM Cyg	56896.3532	34092	0.0801	V	T. Arranz	0.0007
TW Boo	56813.4543	56216	–0.0811	V	T. Arranz	0.0006	DM Cyg	56899.7107	34100	0.0787	V	K. Menzies	0.0007
TW Boo	56817.7129	56224	–0.0807	V	R. Sabo	0.0014	DM Cyg	56912.7280	34131	0.0803	V	R. Sabo	0.0010
TW Boo	56846.4556	56278	–0.0807	V	T. Arranz	0.0007	DM Cyg	56914.8234	34136	0.0764	V	R. Sabo	0.0009
TW Boo	56862.4216	56308	–0.0829	V	T. Arranz	0.0008	DM Cyg	56930.3607	34173	0.0789	V	T. Arranz	0.0005
UU Boo	56750.7411	45229	0.2738	V	K. Menzies	0.0007	DM Cyg	56935.3979	34185	0.0778	V	T. Arranz	0.0006
UU Boo	56828.4198	45399	0.2760	V	T. Arranz	0.0010	DM Cyg	56956.3915	34235	0.0784	V	T. Arranz	0.0008
UU Boo	56838.4722	45421	0.2762	V	T. Arranz	0.0007	DM Cyg	56959.3308	34242	0.0787	V	T. Arranz	0.0007
UU Boo	56865.4308	45480	0.2765	V	T. Arranz	0.0006	RW Dra	56807.7487	39353	0.2450	V	G. Samolyk	0.0014
UY Boo	56766.8033	22940	0.9200	V	G. Samolyk	0.0015	RW Dra	56825.4246	39393	0.2042	V	T. Arranz	0.0008
UY Boo	56782.4261	22964	0.9227	V	T. Arranz	0.0013	RW Dra	56836.5231	39418	0.2298	V	T. Arranz	0.0007
UY Boo	56810.3924	23007	0.9030	V	T. Arranz	0.0007	RW Dra	56848.4956	39445	0.2435	V	T. Arranz	0.0007
UY Cam	56924.7925	79986	–0.0951	V	G. Samolyk	0.0037	RW Dra	56856.4518	39463	0.2272	V	T. Arranz	0.0006
RW Cnc	56676.7445	31287	0.2154	V	R. Sabo	0.0015	RW Dra	56864.4003	39481	0.2032	V	T. Arranz	0.0010
RW Cnc	56682.7602	31298	0.2119	V	R. Sabo	0.0011	RW Dra	56880.3848	39517	0.2427	V	T. Arranz	0.0008
RW Cnc	56725.4536	31376	0.2238	V	T. Arranz	0.0012	RW Dra	56883.4896	39524	0.2471	V	T. Arranz	0.0007
RW Cnc	56730.3699	31385	0.2153	V	T. Arranz	0.0007	RW Dra	56895.4352	39551	0.2339	V	T. Arranz	0.0007
RW Cnc	56731.4685	31387	0.2195	V	T. Arranz	0.0009	RW Dra	56899.4062	39560	0.2187	V	T. Arranz	0.0005
RW Cnc	56736.3869	31396	0.2131	V	T. Arranz	0.0008	RW Dra	56907.3682	39578	0.2082	V	T. Arranz	0.0008
RW Cnc	56742.4046	31407	0.2116	V	T. Arranz	0.0007	SV Eri	56994.7835	30039	0.9444	V	R. Sabo	0.0034
RW Cnc	56743.4979	31409	0.2105	V	T. Arranz	0.0008	BB Eri	56956.8538	30527	0.2866	V	G. Samolyk	0.0013
RW Cnc	56753.3530	31427	0.2160	V	T. Arranz	0.0006	RR Gem	56675.7959	38557	–0.5139	V	R. Sabo	0.0009
RW Cnc	56759.3750	31438	0.2188	V	T. Arranz	0.0008	RR Gem	56691.2925	38596	–0.5124	V	T. Arranz	0.0007
RW Cnc	56760.4680	31440	0.2174	V	T. Arranz	0.0012	RR Gem	56708.3832	38639	–0.5061	V	T. Arranz	0.0011
RW Cnc	56765.3924	31449	0.2170	V	T. Arranz	0.0010	RR Gem	56710.3605	38644	–0.5153	V	T. Arranz	0.0008
RW Cnc	56766.4823	31451	0.2126	V	T. Arranz	0.0011	RR Gem	56712.3461	38649	–0.5163	V	T. Arranz	0.0005
TT Cnc	56773.5965	29868	0.1228	V	G. Samolyk	0.0015	RR Gem	56716.3188	38659	–0.5167	V	T. Arranz	0.0007
TT Cnc	57011.9217	30291	0.1089	V	K. Menzies	0.0012	RR Gem	56723.4730	38677	–0.5141	V	T. Arranz	0.0006
KV Cnc	56687.4848	7821	–0.1504	V	T. Arranz	0.0006	RR Gem	56727.4477	38687	–0.5125	V	T. Arranz	0.0009
KV Cnc	56690.5039	7827	–0.1433	V	T. Arranz	0.0008	RR Gem	56729.4328	38692	–0.5139	V	T. Arranz	0.0008
KV Cnc	56691.5074	7829	–0.1438	V	T. Arranz	0.0005	RR Gem	56733.4000	38702	–0.5198	V	T. Arranz	0.0005
KV Cnc	56774.3623	7994	–0.1189	V	T. Arranz	0.0006	RR Gem	56735.3899	38707	–0.5165	V	T. Arranz	0.0007
KV Cnc	56776.3743	7998	–0.1149	V	T. Arranz	0.0020	RR Gem	56739.3617	38717	–0.5178	V	T. Arranz	0.0004
KV Cnc	56778.3873	8002	–0.1099	V	T. Arranz	0.0010	RR Gem	56741.3464	38722	–0.5197	V	T. Arranz	0.0006
RV Cap	56923.6404	51459	–0.0806	V	G. Samolyk	0.0022	RR Gem	56743.3362	38727	–0.5164	V	T. Arranz	0.0005
RZ Cap	56953.6198	13887	0.0051	V	G. Samolyk	0.0010	RR Gem	56758.4327	38765	–0.5177	V	T. Arranz	0.0007
VW Cap	51410.6407	82737	0.1111	V	G. Samolyk	0.0038	RR Gem	56764.3929	38780	–0.5172	V	T. Arranz	0.0006
VW Cap	54710.8024	93238	0.2285	V	G. Samolyk	0.0017	RR Gem	56766.3801	38785	–0.5165	V	T. Arranz	0.0006
VW Cap	55087.6010	94437	0.2294	V	G. Samolyk	0.0048	RR Gem	56949.9215	39247	–0.5326	V	R. Sabo	0.0009
VW Cap	56892.7148	100181	0.2337	V	G. Samolyk	0.0024	TW Her	56837.8979	88320	–0.0173	V	R. Sabo	0.0009
YZ Cap	56893.6918	48139	0.0380	V	G. Samolyk	0.0022	TW Her	56871.4651	88404	–0.0165	V	T. Arranz	0.0004
RR Cet	56952.7802	42984	0.0146	V	G. Samolyk	0.0013	TW Her	56873.4637	88409	–0.0159	V	T. Arranz	0.0006
RU Cet	56897.8451	29118	0.1331	V	G. Samolyk	0.0017	TW Her	56889.4477	88449	–0.0159	V	T. Arranz	0.0006
RV Cet	56929.8412	28579	0.2439	V	G. Samolyk	0.0017	TW Her	56915.4233	88514	–0.0143	V	T. Arranz	0.0007
RX Cet	57003.5810	29419	0.3429	V	G. Samolyk	0.0018	TW Her	56931.4055	88554	–0.0161	V	T. Arranz	0.0008
UU Cet	56998.6417	26053	–0.1626	R	G. Samolyk	0.0017	VX Her	56791.8505	76951	–0.0261	V	R. Sabo	0.0019
XZ Cyg	56801.8562	27169	–2.3561	V	G. Samolyk	0.0017	VX Her	56807.7860	76986	–0.0286	V	G. Samolyk	0.0008
XZ Cyg	56816.7867	27201	–2.3600	V	K. Menzies	0.0006	VX Her	56828.7307	77032	–0.0311	V	R. Sabo	0.0011
XZ Cyg	56885.3692	27348	–2.3824	V	T. Arranz	0.0007	VX Her	56858.7861	77098	–0.0303	V	R. Sabo	0.0012
XZ Cyg	56892.3660	27363	–2.3861	V	T. Arranz	0.0006	VX Her	56867.4364	77117	–0.0321	V	T. Arranz	0.0006
XZ Cyg	56905.4348	27391	–2.3849	V	T. Arranz	0.0007	VZ Her	56739.7890	45309	0.0746	V	K. Menzies	0.0011
XZ Cyg	56906.3678	27393	–2.3853	V	T. Arranz	0.0007	VZ Her	56813.7656	45477	0.0761	V	R. Sabo	0.0008
XZ Cyg	56912.4398	27406	–2.3804	V	T. Arranz	0.0007	AR Her	56742.9245	32530	–1.4333	V	G. Samolyk	0.0012

Table continued on next page

Table 1. Recent times of maxima of stars in the AAVSO Short Period Pulsator program, cont.

Star	JD (max) Hel. 2400000+	Cycle	O–C	F	Observer	Error	Star	JD (max) Hel. 2400000+	Cycle	O–C	F	Observer	Error
AR Her	56766.9209	32581	−1.4084	V	G. Samolyk	0.0010	SZ Lyn	56952.9421	156208	0.0251	V	G. Samolyk	0.0007
AR Her	56767.8561	32583	−1.4132	V	K. Menzies	0.0006	SZ Lyn	56966.9246	156324	0.0255	V	G. Samolyk	0.0007
AR Her	56807.7835	32668	−1.4382	V	G. Samolyk	0.0011	SZ Lyn	57022.7325	156787	0.0258	V	N. Simmons	0.0005
AR Her	56844.4484	32746	−1.4355	V	T. Arranz	0.0013	SZ Lyn	57022.8531	156788	0.0258	V	N. Simmons	0.0005
AR Her	56846.8362	32751	−1.3978	V	R. Sabo	0.0020	RR Lyr	56915.6843	24684	−0.2988	V	H. Smith	0.0012
AR Her	56851.5440	32761	−1.3903	V	T. Arranz	0.0006	RR Lyr	56919.6504	24691	−0.3008	V	H. Smith	0.0009
AR Her	56852.4934	32763	−1.3810	V	T. Arranz	0.0006	RR Lyr	56940.6201	24728	−0.3052	V	H. Smith	0.0012
AR Her	56853.4178	32765	−1.3966	V	T. Arranz	0.0008	RR Lyr	56953.6547	24751	−0.3085	V	G. Samolyk	0.0014
AR Her	56860.4549	32780	−1.4099	V	T. Arranz	0.0007	RZ Lyr	56805.9235	30558	−0.0447	V	R. Sabo	0.0021
AR Her	56861.3965	32782	−1.4084	V	T. Arranz	0.0010	MW Lyr	56850.6275	52304	0.0348	V	K. Menzies	0.0021
AR Her	56868.4167	32797	−1.4386	V	T. Arranz	0.0008	AV Peg	56807.9068	33346	0.1561	V	R. Sabo	0.0013
AR Her	56876.4430	32814	−1.4028	V	T. Arranz	0.0007	AV Peg	56896.9169	33574	0.1607	V	R. Sabo	0.0014
AR Her	56884.4420	32831	−1.3943	V	T. Arranz	0.0009	AV Peg	56900.8174	33584	0.1575	V	G. Samolyk	0.0011
DL Her	56790.9064	31409	0.0387	V	R. Sabo	0.0017	AV Peg	56902.3821	33588	0.1607	V	T. Arranz	0.0007
DL Her	56881.4331	31562	0.0464	V	T. Arranz	0.0007	AV Peg	56918.3858	33629	0.1590	V	T. Arranz	0.0005
DL Her	56894.4465	31584	0.0440	V	T. Arranz	0.0010	AV Peg	56925.4128	33647	0.1593	V	T. Arranz	0.0006
DL Her	56897.3998	31589	0.0391	V	T. Arranz	0.0007	AV Peg	56936.3449	33675	0.1609	V	T. Arranz	0.0005
DL Her	56898.5882	31591	0.0443	V	K. Menzies	0.0012	AV Peg	56945.3241	33698	0.1615	V	T. Arranz	0.0007
DL Her	56910.4308	31611	0.0543	V	T. Arranz	0.0011	AV Peg	56949.6193	33709	0.1625	V	R. Sabo	0.0010
DY Her	56773.8369	156995	−0.0289	V	G. Samolyk	0.0007	AV Peg	56950.3971	33711	0.1596	V	T. Arranz	0.0009
DY Her	56837.7474	157425	−0.0298	V	R. Sabo	0.0009	AV Peg	56957.4241	33729	0.1598	V	T. Arranz	0.0005
DY Her	56863.4597	157598	−0.0308	V	T. Arranz	0.0004	AV Peg	56997.6339	33832	0.1610	V	R. Sabo	0.0009
LS Her	56806.7692	124787	0.0205	V	G. Samolyk	0.0014	BH Peg	56936.6840	27413	−0.1371	V	R. Sabo	0.0018
SZ Hya	56726.5401	29870	−0.2373	V	T. Arranz	0.0007	BH Peg	57006.5510	27522	−0.1383	V	K. Menzies	0.0018
SZ Hya	56731.3691	29879	−0.2434	V	T. Arranz	0.0007	GV Peg	56887.8397	21114	0.2826	V	K. Menzies	0.0012
SZ Hya	56732.4425	29881	−0.2445	V	T. Arranz	0.0011	RV UMa	56723.7815	24886	0.1293	V	G. Samolyk	0.0016
SZ Hya	56742.5993	29900	−0.2953	V	G. Samolyk	0.0014	RV UMa	56729.8663	24899	0.1294	V	G. Samolyk	0.0009
SZ Hya	56760.3733	29933	−0.2502	V	T. Arranz	0.0010	RV UMa	56737.8255	24916	0.1315	V	G. Samolyk	0.0011
SZ Hya	57022.0069	30420	−0.2526	V	G. Samolyk	0.0009	RV UMa	56746.7167	24935	0.1296	V	G. Samolyk	0.0010
UU Hya	56740.4019	32950	−0.0049	V	T. Arranz	0.0008	RV UMa	56780.4126	25007	0.1252	V	T. Arranz	0.0008
UU Hya	56741.4549	32952	0.0004	V	T. Arranz	0.0006	RV UMa	56786.4995	25020	0.1273	V	T. Arranz	0.0011
UU Hya	56761.3704	32990	0.0089	V	T. Arranz	0.0010	RV UMa	56787.4312	25022	0.1229	V	T. Arranz	0.0009
DG Hya	56729.5799	5774	0.0176	V	G. Samolyk	0.0016	RV UMa	56788.3688	25024	0.1244	V	T. Arranz	0.0008
DH Hya	56742.6300	52283	0.0931	V	G. Samolyk	0.0009	RV UMa	56794.4579	25037	0.1287	V	T. Arranz	0.0013
RR Leo	56773.6914	29793	0.1358	V	G. Samolyk	0.0012	RV UMa	56816.4581	25084	0.1301	V	T. Arranz	0.0007
SS Leo	56779.7467	23946	−0.0981	V	R. Sabo	0.0013	RV UMa	56817.3950	25086	0.1308	V	T. Arranz	0.0006
ST Leo	56753.6763	60317	−0.0207	V	G. Samolyk	0.0011	AE UMa	56660.6296	244792	0.0047	V	G. Samolyk	0.0006
TV Leo	56725.8151	29255	0.1223	V	G. Samolyk	0.0017	AE UMa	56660.7111	244793	0.0002	V	G. Samolyk	0.0006
WW Leo	56663.8393	36131	0.0428	V	R. Sabo	0.0019	AE UMa	56660.7983	244794	0.0013	V	G. Samolyk	0.0008
WW Leo	56695.7932	36184	0.0460	V	K. Menzies	0.0023	AE UMa	56660.8887	244795	0.0057	V	G. Samolyk	0.0010
WW Leo	56750.6512	36275	0.0450	V	K. Menzies	0.0017	AE UMa	56660.9712	244796	0.0022	V	G. Samolyk	0.0006
VY LMi	56739.7118	11883	0.0119	V	K. Menzies	0.0018	AE UMa	56722.4705	245511	−0.0007	V	T. Arranz	0.0004
U Lep	56998.7525	26908	0.0449	R	G. Samolyk	0.0013	AE UMa	56730.4750	245604	0.0042	V	T. Arranz	0.0005
SZ Lyn	56705.4882	154155	0.0294	V	T. Arranz	0.0011	AE UMa	56732.3632	245626	0.0000	V	T. Arranz	0.0003
SZ Lyn	56722.3625	154295	0.0288	V	T. Arranz	0.0006	AE UMa	56744.4049	245766	−0.0006	V	T. Arranz	0.0003
SZ Lyn	56923.8939	155967	0.0258	V	G. Samolyk	0.0013	AE UMa	56998.9327	248725	0.0027	V	G. Samolyk	0.0006
SZ Lyn	56952.8218	156207	0.0253	V	G. Samolyk	0.0007	AE UMa	56999.0159	248726	−0.0001	V	G. Samolyk	0.0008

Recent Minima of 149 Eclipsing Binary Stars

Gerard Samolyk
P.O. Box 20677, Greenfield, WI 53220; gsamolyk@wi.rr.com

Received March 8, 2015; accepted May 12, 2015

Abstract This paper continues the publication of times of minima for 149 eclipsing binary stars from observations reported to the AAVSO EB section. Times of minima from observations received by the author from October 2014 through March 2015 are presented.

1. Recent observations

The accompanying list contains times of minima calculated from recent CCD observations made by participants in the AAVSO's eclipsing binary program. This list will be web-archived and made available through the AAVSO ftp site at ftp://ftp.aavso.org/public/datasets/gsamoj2431.txt. This list, along with the eclipsing binary data from earlier AAVSO publications, is also included in the Lichtenknecker database (Kreiner 2011) administrated by the Bundesdeutsche Arbeitsgemeinschaft für Veränderliche Sterne e. V. (BAV) at: http://www.bav-astro.de/LkDB/index.php?lang=en. These observations were reduced by the observers or the writer using the method of Kwee and van Woerden (1956). The standard error is included when available. Column F indicates the filter used.

The linear elements in the *General Catalogue of Variable Stars* (GCVS; Kholopov *et al.* 1985) were used to compute the O–C values for most stars. For a few exceptions where the GCVS elements are missing or are in significant error, light elements from another source are used: AC CMi (Samolyk 2008), CW Cas (Samolyk 1992a), DF Hya (Samolyk 1992b), EF Ori (Baldwin and Samolyk 2005), GU Ori (Samolyk 1985). The light elements used for QX And, CP Psc, DS Psc, and V1128 Tau are from Kreiner (2004) The light elements used for AG Ari, V610 Aur, V700 Cyg, V2477 Cyg, V2643 Cyg, V740 Per, GR Psc and EQ UMa are from Paschke (2014). O-C values listed in this paper can be directly compared with values published in the AAVSO EB monographs.

References

Baldwin, M. E., and Samolyk, G. 2005, *Observed Minima Timings of Eclipsing Binaries No. 10*, AAVSO, Cambridge, MA.

Kreiner, J. M. 2004, "Up-to-date linear elements of eclipsing binaries," *Acta Astron.*, **54**, 207 (http://www.as.up.krakow.pl/ephem/).

Kholopov, P. N., *et al.* 1985, *General Catalogue of Variable Stars*, 4th ed., Moscow.

Kwee, K. K., and van Worden, H. 1956, *Bull. Astron. Inst. Netherlands*, **12**, 327.

Paschke, A. 2014, "O–C Gateway" (http://var.astro.cz/ocgate/).

Samolyk, G. 1985, *J. Amer. Assoc. Var. Star Obs.*, **14**, 12.

Samolyk, G. 1992a, *J. Amer. Assoc. Var. Star Obs.*, **21**, 34.

Samolyk, G. 1992b, *J. Amer. Assoc. Var. Star Obs.*, **21**, 111.

Samolyk, G. 2008, *J. Amer. Assoc. Var. Star Obs.*, **36**, 171.

Table 1. Recent times of minima of stars in the AAVSO eclipsing binary program.

Star	JD (min) Hel. 2400000+	Cycle	O–C (day)	F	Observer	Error (day)	Star	JD (min) Hel. 2400000+	Cycle	O–C (day)	F	Observer	Error (day)
AB And	56929.6365	62731.5	–0.0352	V	B. Manske	0.0001	SX Aur	56966.7371	13887	0.0179	V	G. Samolyk	0.0002
AB And	56929.8020	62732	–0.0356	V	B. Manske	0.0001	TT Aur	56976.8754	26813	–0.0046	V	K. Menzies	0.0002
AB And	57061.5624	63129	–0.0364	V	V. Petriew	0.0001	AP Aur	56226.9372	24208	1.4112	V	J. A. Howell	0.0011
AD And	56936.8781	18185.5	–0.0372	V	R. Sabo	0.0002	AP Aur	56998.7449	25563.5	1.5158	V	N. Simmons	0.0004
BD And	56952.5922	47505	0.0184	V	G. Samolyk	0.0001	AP Aur	57019.8107	25600.5	1.5171	V	V. Petriew	0.0002
BX And	57003.5633	33559	–0.0751	V	G. Samolyk	0.0001	AR Aur	57020.5911	4503	–0.1237	V	G. Samolyk	0.0001
CN And	57020.6565	33370	–0.1397	V	V. Petriew	0.0001	CL Aur	56210.9024	18679	0.1559	V	J. A. Howell	0.0007
QX And	56943.8572	10781.5	0.0008	V	K. Menzies	0.0004	CL Aur	56953.8004	19276	0.1683	V	G. Samolyk	0.0001
QX And	56999.7063	10917	0.0006	V	K. Menzies	0.0007	EM Aur	56983.6886	14266	–1.1092	V	K. Menzies	0.0004
CX Aqr	56538.8476	36125.5	0.0131	V	B. Manske	0.0005	EP Aur	56983.8913	51902	0.0121	V	K. Menzies	0.0001
CX Aqr	56573.5958	36188	0.0122	V	B. Manske	0.0001	HP Aur	56226.8905	9467.5	0.0590	V	J. A. Howell	0.0031
CX Aqr	56914.6945	36801.5	0.0134	V	B. Manske	0.0002	HP Aur	56956.7992	9980.5	0.0646	V	G. Samolyk	0.0002
OO Aql	56806.7341	35899.5	0.0589	V	B. Manske	0.0001	IM Aur	57074.5297	13276	–0.1185	V	K. Menzies	0.0002
OO Aql	56807.7473	35901.5	0.0585	V	B. Manske	0.0001	V610 Aur	56961.8117	3364	0.0064	V	V. Petriew	0.0005
OO Aql	56860.7071	36006	0.0589	V	B. Manske	0.0002	V610 Aur	57021.7214	3401	0.0060	V	V. Petriew	0.0003
OO Aql	56913.4139	36110	0.0597	V	L. Corp	0.0001	TU Boo	57082.7738	74059.5	–0.1485	V	K. Menzies	0.0001
V343 Aql	56953.5537	15456	–0.0473	V	G. Samolyk	0.0002	TU Boo	57084.8811	74066	–0.1490	V	K. Menzies	0.0001
V346 Aql	56848.7413	13495	–0.0114	V	B. Manske	0.0001	TZ Boo	57054.9189	58628	0.0634	V	K. Menzies	0.0001
SS Ari	56567.8116	43202	–0.3189	V	B. Manske	0.0001	ZZ Boo	57087.8595	3710.5	0.0742	V	G. Samolyk	0.0001
SS Ari	56908.8333	44042	–0.3318	V	B. Manske	0.0003	AD Boo	57081.8709	15127	0.0329	V	K. Menzies	0.0001
SS Ari	56998.5543	44263	–0.3354	V	N. Simmons	0.0001	AD Boo	57082.9045	15128	0.0321	V	K. Menzies	0.0003
AG Ari	57014.3001	4337	–0.0017	V	L. Corp	0.0003	Y Cam	56260.8679	4023	0.4134	V	J. A. Howell	0.0010

Table continued on following pages

Table 1. Recent times of minima of stars in the AAVSO eclipsing binary program, cont.

Star	JD (min) Hel. 2400000+	Cycle	O–C (day)	F	Observer	Error (day)	Star	JD (min) Hel. 2400000+	Cycle	O–C (day)	F	Observer	Error (day)
Y Cam	57090.6107	4274	0.4444	V	G. Samolyk	0.0001	YY Del	56527.7430	17108	0.0096	V	B. Manske	0.0002
AL Cam	57090.6786	23096	–0.0315	V	N. Simmons	0.0001	YY Del	56903.6698	17582	0.0107	V	B. Manske	0.0001
RT CMa	56263.8334	22909	–0.7363	V	J. A. Howell	0.0004	FZ Del	56917.7246	32677.5	–0.0341	V	B. Manske	0.0004
UU CMa	57081.6053	5762	–0.0887	V	G. Samolyk	0.0001	S Equ	56966.5666	4182	0.0659	V	G. Samolyk	0.0001
XZ CMi	57090.5965	25304	–0.0008	V	N. Simmons	0.0001	TX Gem	56956.9598	13253	–0.0377	V	G. Samolyk	0.0001
YY CMi	57095.6437	26574	0.0172	V	G. Samolyk	0.0001	WW Gem	57096.6739	25135	0.0374	V	G. Samolyk	0.0002
AC CMi	57006.8760	5798	0.0032	V	K. Menzies	0.0001	AF Gem	57089.6387	24067	–0.0691	V	G. Samolyk	0.0001
AK CMi	56225.9434	23192	–0.0234	V	J. A. Howell	0.0011	TU Her	57082.8816	5744	–0.2313	V	K. Menzies	0.0001
AK CMi	57019.8996	24595	–0.0214	V	G. Samolyk	0.0001	UX Her	57093.9343	11248	0.1146	V	G. Samolyk	0.0002
AM CMi	57081.7087	31237.5	0.2330	V	G. Samolyk	0.0007	HS Her	56769.8283	7090	–0.0199	V	K. Menzies	0.0008
TV Cas	57058.5827	6872	–0.0277	V	G. Lubcke	0.0002	DF Hya	56966.9384	42888.5	–0.0002	V	G. Samolyk	0.0003
TV Cas	57058.5827	6872	–0.0277	Ic	G. Lubcke	0.0003	DF Hya	57093.5597	43271.5	–0.0007	V	G. Samolyk	0.0001
TV Cas	57058.5830	6872	–0.0273	B	G. Lubcke	0.0002	RT Lac	56913.5380	2373	–0.3092	V	L. Corp	0.0002
ZZ Cas	56958.8207	18915	0.0125	V	K. Menzies	0.0004	SW Lac	56948.3726	36396.5	–0.0933	V	L. Corp	0.0002
CW Cas	56557.6561	46808.5	–0.0791	V	B. Manske	0.0004	SW Lac	56948.5360	36397	–0.0903	V	L. Corp	0.0003
MM Cas	57003.5820	18647	0.1089	V	G. Samolyk	0.0002	SW Lac	56999.5304	36556	–0.0905	V	K. Menzies	0.0001
OR Cas	56977.6589	10249	–0.0283	V	K. Menzies	0.0001	CO Lac	57036.5057	19130	0.0034	V	K. Menzies	0.0001
OR Cas	56982.6415	10253	–0.0285	V	N. Simmons	0.0001	Y Leo	57019.9198	6870	–0.0519	V	G. Samolyk	0.0001
OX Cas	56998.5823	6318.5	0.0350	V	G. Samolyk	0.0005	UU Leo	57093.6860	6963	0.1941	V	G. Samolyk	0.0001
PV Cas	56998.6213	9581	–0.0348	V	G. Samolyk	0.0001	UV Leo	57110.6051	31112	0.0411	V	N. Simmons	0.0001
DK Cep	56956.5769	23700	0.0322	V	N. Simmons	0.0001	VZ Leo	57017.9172	23721	–0.0590	V	K. Menzies	0.0001
SS Cet	57053.6105	4910	0.0593	V	G. Samolyk	0.0003	AM Leo	57081.7678	39881	0.0127	V	K. Menzies	0.0001
TX Cet	56966.7303	18741	0.0089	V	G. Samolyk	0.0004	Z Lep	56232.9230	28991	–0.1796	V	J. A. Howell	0.0007
RW Com	57105.6293	71976	0.0045	V	K. Menzies	0.0001	Z Lep	57080.5545	29844	–0.1870	V	G. Samolyk	0.0001
SS Com	56766.6406	76947.5	0.8259	V	N. Simmons	0.0002	RR Lep	56238.8812	28251	–0.0372	V	J. A. Howell	0.0010
SS Com	57084.9305	77718.5	0.8532	V	K. Menzies	0.0003	RY Lyn	56229.9290	9298	–0.0324	V	J. A. Howell	0.0005
TW CrB	57096.8669	32602	0.0525	V	N. Simmons	0.0001	RY Lyn	57017.7426	9847	–0.0272	V	K. Menzies	0.0001
W Crv	57081.9256	44924	0.0164	V	G. Samolyk	0.0001	UV Lyn	56993.0396	40294.5	0.0893	V	V. Petriew	0.0001
V Crt	57080.8134	22340	–0.0054	V	G. Samolyk	0.0001	UV Lyn	57004.8672	40323	0.0900	V	V. Petriew	0.0002
WW Cyg	56943.6279	4993	0.1213	V	K. Menzies	0.0001	UV Lyn	57020.8444	40361.5	0.0905	V	V. Petriew	0.0002
ZZ Cyg	56956.5754	19020	–0.0663	V	G. Samolyk	0.0001	UV Lyn	57021.0518	40362	0.0903	V	V. Petriew	0.0002
BR Cyg	56966.5620	11577	0.0014	V	G. Samolyk	0.0001	EW Lyr	56956.5492	15629	0.2604	V	G. Samolyk	0.0001
GO Cyg	56950.5920	32072	0.0652	V	V. Petriew	0.0003	Beta Lyr	56826.16	605.5	1.76	V	G. Samolyk	0.06
V401 Cyg	56484.7720	21707.5	0.0752	Ic	G. Lubcke	0.0002	Beta Lyr	56826.18	605.5	1.78	R	G. Samolyk	0.06
V401 Cyg	56484.7722	21707.5	0.0754	V	G. Lubcke	0.0002	Beta Lyr	56826.20	605.5	1.80	B	G. Samolyk	0.02
V401 Cyg	56484.7722	21707.5	0.0754	B	G. Lubcke	0.0001	Beta Lyr	56832.69	606	1.82	R	G. Samolyk	0.05
V401 Cyg	56485.6434	21709	0.0725	B	G. Lubcke	0.0005	Beta Lyr	56832.71	606	1.84	V	G. Samolyk	0.05
V401 Cyg	56485.6444	21709	0.0735	V	G. Lubcke	0.0006	Beta Lyr	56832.72	606	1.85	B	G. Samolyk	0.03
V401 Cyg	56485.6446	21709	0.0737	Ic	G. Lubcke	0.0007	RU Mon	57090.6550	4281.5	–0.6425	V	G. Samolyk	0.0001
V401 Cyg	56538.6725	21800	0.0739	B	G. Lubcke	0.0001	RW Mon	57079.5780	12276	–0.0815	V	K. Menzies	0.0001
V401 Cyg	56538.6731	21800	0.0745	V	G. Lubcke	0.0002	BB Mon	56966.9312	41334	–0.0038	V	G. Samolyk	0.0003
V401 Cyg	56538.6740	21800	0.0754	Ic	G. Lubcke	0.0002	BO Mon	56266.9598	5734	–0.0447	V	J. A. Howell	0.0004
V401 Cyg	56557.6140	21832.5	0.0769	V	G. Lubcke	0.0002	EP Mon	56976.8769	20981	0.0295	V	N. Simmons	0.0003
V401 Cyg	56557.6141	21832.5	0.0770	Ic	G. Lubcke	0.0001	V508 Oph	56799.7632	33983.5	–0.0231	V	B. Manske	0.0001
V401 Cyg	56557.6147	21832.5	0.0776	B	G. Lubcke	0.0002	V839 Oph	56800.7384	39981	0.2836	V	B. Manske	0.0001
V401 Cyg	56559.6503	21836	0.0737	B	G. Lubcke	0.0004	EF Ori	56272.8999	2422	0.0063	V	J. A. Howell	0.0006
V401 Cyg	56559.6513	21836	0.0747	Ic	G. Lubcke	0.0003	EQ Ori	56230.9664	14199	–0.0399	V	J. A. Howell	0.0008
V401 Cyg	56559.6516	21836	0.0750	V	G. Lubcke	0.0003	ER Ori	56242.8766	34521.5	0.1034	V	J. A. Howell	0.0014
V401 Cyg	56561.6928	21839.5	0.0767	V	G. Lubcke	0.0002	ER Ori	56956.9540	36208	0.1189	V	G. Samolyk	0.0002
V401 Cyg	56561.6930	21839.5	0.0769	B	G. Lubcke	0.0002	ET Ori	56208.9264	31048	–0.0051	V	J. A. Howell	0.0005
V401 Cyg	56561.6933	21839.5	0.0772	Ic	G. Lubcke	0.0002	ET Ori	57087.5937	31972	–0.0023	V	G. Samolyk	0.0002
V401 Cyg	56566.6430	21848	0.0737	B	G. Lubcke	0.0004	FH Ori	57076.7173	14493	–0.4316	V	R. Sabo	0.0002
V401 Cyg	56566.6438	21848	0.0745	V	G. Lubcke	0.0003	FL Ori	56222.9134	7012	0.0407	V	J. A. Howell	0.0001
V401 Cyg	56566.6449	21848	0.0756	Ic	G. Lubcke	0.0002	FL Ori	57080.5979	7565	0.0383	V	G. Samolyk	0.0002
V401 Cyg	56848.6886	22332	0.0819	B	G. Lubcke	0.0003	FR Ori	56962.8389	32950.5	0.0448	V	B. Manske	0.0007
V401 Cyg	56848.6895	22332	0.0828	V	G. Lubcke	0.0001	FT Ori	56243.8629	4728	0.0176	V	J. A. Howell	0.0010
V401 Cyg	56848.6904	22332	0.0837	Ic	G. Lubcke	0.0002	FT Ori	56977.9111	4961	0.0191	V	K. Menzies	0.0001
V401 Cyg	56929.6894	22471	0.0843	B	G. Lubcke	0.0002	FZ Ori	56258.7999	30587	–0.0485	V	J. A. Howell	0.0023
V401 Cyg	56929.6895	22471	0.0844	V	G. Lubcke	0.0002	FZ Ori	56976.7799	32082	–0.0445	V	N. Simmons	0.0002
V401 Cyg	56929.6904	22471	0.0853	Ic	G. Lubcke	0.0001	FZ Ori	56998.7800	32437	–0.0436	V	G. Samolyk	0.0005
V456 Cyg	56966.5601	13527	0.0502	V	G. Samolyk	0.0001	GU Ori	56272.9218	28051	–0.0539	V	J. A. Howell	0.0004
V680 Cyg	56948.5825	10948	0.0682	V	V. Petriew	0.0009	GU Ori	56956.8175	29504	–0.0577	V	B. Manske	0.0004
V700 Cyg	56951.5892	85754	–0.0073	V	V. Petriew	0.0001	U Peg	56952.6547	54541.5	–0.1554	V	B. Manske	0.0002
V1425 Cyg	57002.5984	13256	0.0124	V	V. Petriew	0.0004	U Peg	56957.3392	54554	–0.1557	R	L. Corp	0.0003
V2477 Cyg	57021.5711	16011.5	0.0000	V	V. Petriew	0.0001	U Peg	56976.6408	54605.5	–0.1553	V	K. Menzies	0.0003
V2643 Cyg	56961.6335	9200	–0.0057	V	V. Petriew	0.0001	BB Peg	56521.7325	35290	–0.0100	V	B. Manske	0.0002

Table continued on next page

Table 1. Recent times of minima of stars in the AAVSO eclipsing binary program, cont.

Star	JD (min) Hel. 2400000+	Cycle	O–C (day)	F	Observer	Error (day)	Star	JD (min) Hel. 2400000+	Cycle	O–C (day)	F	Observer	Error (day)
BB Peg	56857.7462	36219.5	−0.0125	V	B. Manske	0.0002	WY Tau	57006.6894	28299	0.0614	V	K. Menzies	0.0001
BB Peg	56882.6902	36288.5	−0.0122	V	B. Manske	0.0001	AC Tau	56206.9648	5173	0.0942	V	J. A. Howell	0.0014
BB Peg	56882.8695	36289	−0.0136	V	B. Manske	0.0002	AM Tau	56214.9313	5363	−0.0608	V	J. A. Howell	0.0005
BG Peg	57006.4733	5877	−2.1801	V	K. Menzies	0.0006	AM Tau	56999.7945	5747	−0.0652	V	K. Menzies	0.0001
DI Peg	56953.5685	16517	0.0024	V	N. Simmons	0.0001	CT Tau	56943.7954	17305	−0.0619	V	V. Petriew	0.0002
GP Peg	56567.6741	15712.5	−0.0502	V	B. Manske	0.0007	CT Tau	56983.8046	17365	−0.0626	V	K. Menzies	0.0001
GP Peg	56908.6515	16062	−0.0512	V	B. Manske	0.0001	EQ Tau	56942.7831	49010	−0.0309	V	B. Manske	0.0002
GP Peg	56927.6757	16081.5	−0.0516	V	B. Manske	0.0002	EQ Tau	56956.7785	49051	−0.0308	V	G. Samolyk	0.0001
Z Per	56236.8658	3461	−0.2563	V	J. A. Howell	0.0006	EQ Tau	56976.7478	49109.5	−0.0304	V	K. Menzies	0.0001
RT Per	56253.8861	26934	0.0796	V	J. A. Howell	0.0004	V1128 Tau	57072.3195	14972.5	−0.0009	V	L. Corp	0.0002
RT Per	56953.8052	27758	0.0928	V	G. Samolyk	0.0001	V Tri	56205.9088	54223	−0.0049	V	J. A. Howell	0.0004
ST Per	56231.8853	5209	0.3045	V	J. A. Howell	0.0004	V Tri	56578.6839	54860	−0.0058	V	B. Manske	0.0001
XZ Per	56262.9126	11076	−0.0643	V	J. A. Howell	0.0002	V Tri	56902.8873	55414	−0.0064	V	R. Sabo	0.0001
XZ Per	56953.8846	11676	−0.0728	V	G. Samolyk	0.0001	V Tri	56949.7038	55494	−0.0063	V	V. Petriew	0.0001
DM Per	56955.7676	5512	−0.0044	V	V. Petriew	0.0005	V Tri	56952.6297	55499	−0.0064	V	N. Simmons	0.0001
IQ Per	57002.7140	7291	−0.0017	V	V. Petriew	0.0002	V Tri	56966.6747	55523	−0.0064	V	B. Manske	0.0001
IU Per	56956.7402	13238	0.0059	V	N. Simmons	0.0001	RV Tri	57003.6374	14556	−0.0399	V	N. Simmons	0.0001
IU Per	56998.7339	13287	0.0054	V	G. Samolyk	0.0002	W UMa	57093.6424	33953	−0.0898	V	G. Samolyk	0.0001
V432 Per	56951.7861	65556	0.0417	V	V. Petriew	0.0001	W UMa	57093.8103	33953.5	−0.0887	V	G. Samolyk	0.0001
V432 Per	56951.9797	65556.5	0.0745	V	V. Petriew	0.0003	TY UMa	57080.5651	49494.5	0.3579	V	G. Samolyk	0.0001
V432 Per	56966.7350	65602.5	0.0400	V	G. Samolyk	0.0001	TY UMa	57080.7424	49495	0.3579	V	G. Samolyk	0.0001
V432 Per	56982.8337	65652.5	0.0629	V	K. Menzies	0.0001	TY UMa	57080.9200	49495.5	0.3582	V	G. Samolyk	0.0002
V432 Per	56999.6991	65705	0.0486	V	K. Menzies	0.0002	UX UMa	57081.6559	99907	−0.0021	V	G. Samolyk	0.0001
V740 Per	56944.7603	14869	0.0038	V	V. Petriew	0.0001	UX UMa	57081.8533	99908	−0.0013	V	G. Samolyk	0.0001
V740 Per	56950.7292	14885	0.0040	V	V. Petriew	0.0001	UX UMa	57084.8034	99923	−0.0013	V	K. Menzies	0.0001
V740 Per	56950.9158	14885.5	0.0041	V	V. Petriew	0.0001	VV UMa	56746.6843	15903	−0.0564	Ic	G. Lubcke	0.0001
Y Psc	56962.6524	3008	−0.0157	V	B. Manske	0.0001	VV UMa	56746.6843	15903	−0.0564	V	G. Lubcke	0.0001
RV Psc	56927.8651	58749	−0.0586	V	B. Manske	0.0002	VV UMa	56746.6845	15903	−0.0562	B	G. Lubcke	0.0001
RV Psc	56982.7101	58848	−0.0587	V	K. Menzies	0.0002	VV UMa	57003.7545	16277	−0.0663	V	G. Samolyk	0.0001
SX Psc	56903.8276	13212	0.0014	V	B. Manske	0.0002	ZZ UMa	56998.9086	9154	−0.0014	V	G. Samolyk	0.0002
SX Psc	56956.6835	13276	0.0010	V	B. Manske	0.0002	EQ UMa	53496.6794	13374	0.0119	V	V. Petriew	0.0004
UV Psc	56962.4148	15743.5	−0.0200	V	L. Corp	0.0009	EQ UMa	53746.8453	14072.5	0.0077	V	V. Petriew	0.0010
CP Psc	56917.8734	6458	0.0004	V	B. Manske	0.0003	EQ UMa	53779.7923	14164.5	0.0046	V	V. Petriew	0.0008
DS Psc	56561.8335	11859	−0.0010	V	B. Manske	0.0008	EQ UMa	53779.9694	14165	0.0026	V	V. Petriew	0.0005
DS Psc	56923.8445	12916	−0.0039	V	B. Manske	0.0004	RU UMi	57092.6677	29521	−0.0146	V	N. Simmons	0.0001
GR Psc	56558.8218	10299	0.0038	V	B. Manske	0.0002	AG Vir	57095.8752	18149	−0.0079	V	G. Samolyk	0.0003
GR Psc	56952.7947	11096	0.0029	V	B. Manske	0.0002	AH Vir	57093.8187	27677.5	0.2712	V	G. Samolyk	0.0001
UZ Pup	57047.9416	15643.5	−0.0114	V	H. Pavlov	0.0003	AW Vir	56801.7443	33274.5	0.0278	V	B. Manske	0.0001
UZ Pup	57093.6472	15701	−0.0098	V	G. Samolyk	0.0001	AW Vir	57081.9330	34066	0.0279	V	G. Samolyk	0.0001
Y Sex	57079.7354	36476	−0.0091	V	K. Menzies	0.0002	AZ Vir	56808.6990	36698.5	−0.0260	V	B. Manske	0.0001
RW Tau	56573.7756	3933	−0.2588	V	B. Manske	0.0001	AZ Vir	57096.8251	37522.5	−0.0240	V	G. Samolyk	0.0002
RZ Tau	56209.9076	44586	0.0684	V	J. A. Howell	0.0010	AZ Vir	57104.8674	37545.5	−0.024	V	K. Menzies	0.0001
RZ Tau	57054.5664	46618	0.0762	V	K. Menzies	0.0001	BT Vul	56952.6052	18884	0.0044	V	G. Samolyk	0.0002
TY Tau	57096.6097	33326	0.2653	V	G. Samolyk	0.0002	CD Vul	56952.6187	15582	−0.0009	V	G. Samolyk	0.0001
WY Tau	56250.8868	27208	0.0583	V	J. A. Howell	0.0015							

Video Technique for Observing Eclipsing Binary Stars

Hristo Pavlov
9 Chad Place, St. Clair, NSW 2759, Australia; hristo_dpavlov@yahoo.com

Anthony Mallama
14012 Lancaster Lane, Bowie, MD 20715; anthony.mallama@gmail.com

Received February 9, 2015; revised March 10, 2015; accepted May 12, 2015

Abstract Video recording has been used for more than a decade to time astronomical events such as stellar occultations. We present a technique for using video to determine the time of minimum of eclipsing binary stars and we examine various aspects of using video. The free open source software packages OCCUREC and TANGRA have been enhanced to offer better support for the recording and reduction of video observations of eclipsing binaries. We present our work in a style and detail that is appropriate for both video observers unfamiliar with variable stars and for variable star observers unfamiliar with video. We present the results of ten times of minima of southern eclipsing binary stars determined using the video technique.

1. Introduction

Analogue video cameras using CCDs with on-chip microlenses are widely used by amateurs to observe stellar occultations, meteors, planets, lunar impacts, asteroids, and comets (Mousis *et al.* 2014). The main advantage of video in those cases is the precise timestamps associated with each video field. Timing devices such as IOTA-VTI (VideoTimers 2011) have access to atomic time reference provided by satellites (GPS time, for example) and can timestamp each video field with a precision better than 1 millisecond (see Figure 1). The photometric data along with the precise timestamp can then be used to build light curves of the observed objects for the duration of the observation. Because of the nature of video there are no gaps in the light curves caused by dead time but there is more noise.

The video observations conducted by amateurs are used in various projects driven by the professionals. An example of such a collaboration is the PHEMU09 campaign for observing mutual events of the satellites of Jupiter in 2009 which was organized by the Institut de Mecanique Celeste et de Calcul des Ephemerides in Paris. In this project 43% of the observations were recorded using a video camera and 57% were done with a CCD camera (Arlot *et al.* 2014). The photometric observations were used to derive astrometric positions of the satellites. Theoretical light curve models were fitted to the observed light curves. In many cases "light curves were perfectly modelled to noisy video observations" (Arlot *et al.* 2014). An example of such a light curve is shown in Figure 2.

Because video had been used very successfully in other astronomical projects, we decided to develop a video technique for observing eclipsing binaries that is accessible to the large group of amateur video observers and to put this technique to a test.

Before we describe the video technique and our results we cover some basics related to analogue video, video cameras, and using video for photometry.

2. Analogue video

There are two major analogue video formats, PAL and NTSC. NTSC is primarily used in the United States. It has a frame resolution of 720 × 480 and runs at 29.97 frames per second (fps). PAL has a frame resolution of 720 × 576 pixels, runs at 25 fps, and is mostly used in Europe and Australasia.

Figure 1. A typical analogue video frame with an IOTA-VTI timestamp inserted at the bottom. The first two characters show the status of the GPS fix, which in this case means a P fix with 9 tracked satellites. Because this is interlaced video the last digit of the OSD is blurred as it is a 6 interleaving with a 7.

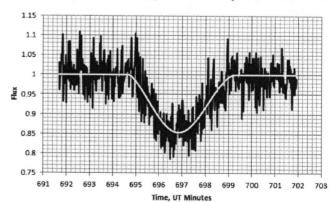

Figure 2. An example of how noisy video data are used very successfully to fit light curve models. This plot shows the video data and the fitted model of J1 occulting J3 on 8 May 2009 (NSDC 2014).

According to the standard each video frame consists of two interleaving half-height video fields which are sent sequentially. Because of this the CCD chip of analogue video cameras is read differently—first the odd lines are read and sent as a video field and then the even lines are read and sent as another video field. In order to reconstruct the video frame the horizontal lines from each video field are alternated to build up the full-height image.

As the analogue video standard has been developed for the images to be watched on a TV screen, a so-called Gamma correction can be applied to the produced video frames. This is an exponential per-pixel correction which will modify the image in such a way that when it is displayed by the TV screen, the brightness of the displayed objects will match closely that of the recorded objects because Gamma will reverse the non-linearity of the TV. In the old days Gamma was always applied unconditionally. In astronomy the analogue video frames will be recorded and displayed in a digital format; a Gamma of 1.00 should be applied when observations are to be used for photometry. Modern video cameras used in astronomy typically allow Gamma to be controlled and to be turned off completely by setting it to 1.00. This guarantees that the resulting system response due to off-sensor processing will not deviate from linearity. The sensor itself may still have a non-linear response, and should be tested, but modern video cameras use recently-developed CCD chips which are expected to have a largely linear response.

The other important difference of analogue video is the bit depth of the pixels—8-bit, compared to the typical 16-bit depth produced by CCD cameras. While this may suggest that video offers significantly lower photometric resolution, in reality this is not exactly the case. This is because video observers adjust the linear Gain of the video camera until the stars appear bright enough but not saturated. This means that video records will usually use most or all of the 8 bits. In contrast, CCD cameras may use a lot less than the 16 bits they support, particularly with shorter exposures that do not lead to levels close to saturation. We can say that while CCD cameras will remain superior when it comes to pixel bit depths and small noise in the domain of shorter exposures, video can produce images with a dynamic range that is comparable to that of a CCD camera running in a short exposure regime.

3. Integrating video cameras

The typical video frame exposure is 40 ms for PAL and 33.37 ms for NTSC, which is insufficient for faint objects. Video cameras perform long exposures by integrating. While the exposure needs to be increased so more light is collected by the CCD chip, at the same time the camera must conform to the analogue video standard by sending video fields with the corresponding constant frame rate. Different models of video cameras may implement this slightly differently, but, generally speaking, all of them will use some sort of an internal buffer to copy the latest long exposure to and then will send this image with the required frame rate while taking another long exposure. Because there must be no dead time and all video frames are sent with the same frequency, the exposures of the video cameras, called integration rates, are a whole number of standard video frame durations and they typically double; for example, ×2, ×4, ×8, ×16, ×32, ×64, ×128, and ×256. In PAL the ×4 integration rate will correspond to a 160 ms exposure. There are two issues to be dealt with as a result of the operation of integrating video cameras.

The first issue is that the integrated image is sent after it has been taken and the timestamp associated with it, inserted during the transmission of the individual video fields, will be off by a constant amount that depends on the camera model and the integration rate which was used. Also, all individual video fields will be time stamped even though they are different fields representing the same long exposure. All this complicates the determination of the time of the exposure. This problem is solved, however, by experimentally determining the so-called Instrumental Delay of each video camera model and each supported exposure and then applying a correction to derive the exact time of the mid-exposure. Most popular video cameras used for astronomy have been tested and their Instrumental Delays are well known (Dangl 2008). This allows photometric data reduced from video to be timed with a precision of milliseconds from UTC.

The second issue to be addressed is that the video produced by integrating cameras will still run at a constant high frame rate even though the exposure is longer. If a standard video recording software is used, a two-hour constant monitoring of a star could produce a file size of 100 to 200 gigabytes. This could create a storage problem as well as data reduction difficulty due to the file size. Also, there is a need to deal with the many repeated video frames of the same image during the data reduction. In order to address those issues we use the custom-built video recording software, OCCUREC, which is described next.

4. Recording software

OCCUREC (Pavlov 2014) is open source software for WINDOWS that allows multiple video frames to be stacked on the fly so the effective frame rate of the video is reduced by saving only one video frame from a physical exposure or combining multiple exposures into one single video frame (Pavlov 2014). This can significantly decrease the recorded file size, making it easier to record video for hours, and it also improves the signal-to-noise ratio (SNR) as a result of the stacking. A new Astro Analogue Video (AAV) open source and lossless file format is used to save the video. AAV is an extension of the Astronomical Digital Video (ADV) file format (Barry *et al.* 2013) customized for analogue video. Because multiple 8-bit video frames are stacked and saved as one, the resulting AAV video bit depth will be more than 8 bits—up to 12 bits with the typical number of frames being stacked.

In our observations of eclipsing binary stars we use OCCUREC as the video recording software. Apart from the convenience of scheduled recording and telescope and video camera control, it offers better SNR and bit depth with a much smaller file size.

The stars that we observed were in the range from magnitude 9 to 12. This was bright for the 14-inch telescope and the very sensitive WAT-910BD video camera so no video integration was required. Because of this OCCUREC was run in a stacking mode in which it combines multiple exposures into a single video frame. A stacking of ×8, ×16, and ×32 was used with the PAL camera, which produced effective exposures of 320 ms, 640 ms, and 1,280 ms.

5. Reduction software

The data reduction has two distinct parts. The first is reducing photometric data from the video and producing a light curve, and the second is analyzing the light curve data to derive a time of minimum (TOM) for the observed star.

To reduce photometric data from the video we use the TANGRA software package (Pavlov 2014). It works with AAV files and has been used previously for video data reduction by a number of researchers (Kitting et al. 2012; Sposetti et al. 2012; Lena et al. 2014). TANGRA offers a range of measurement options including aperture photometry, PSF photometry, and various ways to measure the background. For analogue video where the smaller bit depth may result in a lot fewer unique light levels for a pixel, we recommend using Aperture Photometry with Average Background.

TANGRA offers an open add-in based integration which allowed us to build an add-in module that implements the standard analysis algorithm of Kwee and van Woerden (1956) to derive a time of minimum and an estimate of its uncertainty from the light curve data (Mallama and Pavlov 2015). The add-in automatically applies the correction for the Heliocentric Julian Day (HJD) from the determined JD of the minimum and reports the TOM in HJD.

TANGRA and OCCUREC have been traditionally used for occultations and to time shorter events with higher precision. During the development and testing of the video technique for observing eclipsing binaries, a number of improvements were made in both software packages to ensure the smooth operation and ease of processing of observations of those stars.

6. Observational guidelines

In order to determine the TOM from a single video observation the record has to be long enough to contain a large number of data points during both the descending and ascending branches of the light curve. If the star exhibits a total eclipse with a flat bottom in the light curve the record has to be even longer. Stars with periods less than one day are typically better suited for video observations as they usually show faster change in brightness. For them, a recording of one to two hours on each side of the minimum/flat bottom is sufficiently long.

During such a long observation the light from the target and comparison stars will pass through varying air masses as their altitude changes. As a result, if there is a large difference in the color of the target and comparison stars there will be a systematic error during the normalization which could affect the determination of the TOM. While we recommend the use of a photometric filter, this is not strictly required. "Taking unfiltered images is usually OK when the target is more than 30 degrees above the horizon or when the variable and comparison stars are the same color. Trouble could happen with unfiltered images of stars close to the horizon." (Samolyk 2015).

7. Observational results

Video observations using the described technique were done at Tangra Observatory (IAU Code E24) by one of the authors (HP; AAVSO code PHRA) using a 14-inch LX-200 ACF and WAT-910BD TACOS video system (Gault 2014). Most observations were done with a Sloan r' filter. Ten times of minima of eight southern stars were determined and are given in Table 1. We used observations and O–C data from (Nelson 2014) and (Paschke 2015) to verify our measurements and to refine the ephemeris of one of the stars we observed.

CG Pup is a semi-detached Algol-type binary that ranges from photographic magnitude 10.2 to 13.0 according to the AAVSO International Variable Star Index (VSX; Watson et al. 2014). The variability was discovered by Hoffmeister (1943) and TOMs prior to the present study were derived from photographic observations.

One of the two video light curves for CG Pup is shown in Figure 3. The O–C values for the ephemeris in the *General Catalogue of Variable Stars* (GCVS; Kholopov et al. 1985) for the video TOMs were 0.38185 and 0.38171 day, which are more than one-third of the orbital period. The difference between the two O–C values, 0.00016 day, matches well with the reported TOM uncertainty of 0.0001.

Table 1. Times of minima of southern stars determined using the video technique in the period December 2014 to February 2015, in addition to one value derived from ASAS photometry.

Star	Type	Time of Minima (HJD)	Uncertainty	Comment
WZ Ant	I	2457061.9718	0.0006	r'
GW Car	I	2457037.1644	0.0004	r'
V576 Cen	I	2457057.1027	0.0003	r'
FT Lup	I	2457061.1252	0.0002	r'
ER Ori	I	2457011.9964	0.0003	clear
ER Ori	I	2457025.9693	0.0002	r'
CG Pup	I	2457036.9680	0.0001	r'
CG Pup	I	2457039.0670	0.0001	r'
CG Pup	I	2453286.8054		ASAS data
UX Ret	I	2457009.9654	0.0003	clear
DS Vel	I	2457029.9710	0.0003	r'

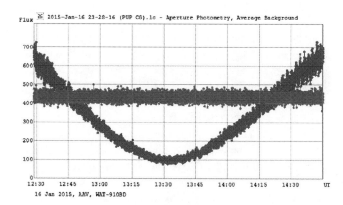

Figure 3. A TANGRA light curve of the minimum of CG Pup that contains more than 12,000 data points. The observation is done with a SLOAN r' filter and the light curve is very symmetrical. The curved band represents the variable and the horizontal band the comparison star.

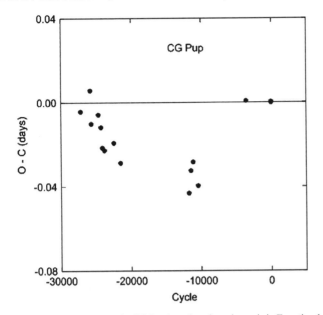

Figure 4. The O–C diagram for CG Pup based on the ephemeris in Equation 1. The two video TOMs are represented by the symbol furthest to the right. Since their O–C difference was only 0.00016 day the two plotted symbols overlap.

The most recently observed TOM of CG Pup listed by Nelson (2014) was in 1985. In order to fill the thirty-year gap we derived the new TOM listed in Table 1 from ASAS data (Pojmański 1997). Our ephemeris in equation 1 is based on the epoch of the most recent video TOM and the best fitting period for the entire 72 years of CG Pup data.

$$HJD = 2457039.0670 + 1.04958387 \times N, \quad (1)$$

where N is the number of cycles.

The O–Cs plotted in Figure 4 suggest that at least one period change has occurred. The residuals show a strong negative slope between cycles –30,000 and –20,000 (years 1943 through 1953) and a minimum around cycle –11,500 (1982). Nevertheless, the ASAS (cycle –3,574) minimum and the video minimum (cycle 0) fit the updated ephemeris very well.

8. Conclusions

We developed a technique for observing eclipsing binaries using video and for determining the time of minimum (TOM) from the data. The tests showed a good agreement of the derived TOM with historical observations, with the Kwee and van Worden (1956) reduction algorithm giving error bars from 0.0001 to 0.0006 day. We derived two TOMs for CG Pup and for ER Ori, which allowed us to compare the O–C value difference. The difference was 0.00016 and 0.00074, respectively, which is comparable to that achievable with a CCD camera.

9. Acknowledgements

The planning of the observations relied heavily on the meteorological forecast data provided by 7Timer (Ye 2011).

This research used information from the International Variable Star Index (VSX) database, operated at AAVSO, Cambridge, Massachusetts, USA.

References

Arlot, J.-E., et al. 2014, *Astron. Astrophys.*, **572A**, 120.

Barry, T., Gault, D., and Pavlov, H. 2013, Astronomical Digital Video System (http://www.kuriwaobservatory.com/ADVS.html).

Dangl, G. 2008, Video Exposure Time Analysis (http://www.dangl.at/ausruest/vid_tim/vid_tim1.htm).

Gault, D., et al. 2014, The Watec WAT-910BD TACOS System (http://www.kuriwaobservatory.com/TACOS_BD-System.html).

Hoffmeister, C. 1943, Kleine Veröff. Berlin-Babelsberg, No. 27, 1.

Kholopov, P. N., et al. 1985, *General Catalogue of Variable Stars*, 4th ed., Moscow.

Kitting, C., Nolthenius, R., Bellerose, J., and Jenniskens, P. 2012, *J. Amer. Inst. Aeronaut. Astronaut.*, **2012**, 1278.

Kwee, K. K., and van Woerden, H. 1956, *Bull. Astron. Inst. Netherlands*, **12**, 327.

Lena, R., Manna, A., and Sposetti, S. 2014, *Lunar Planet. Inst. Contrib.*, No. 1777, 1009.

Mallama, A., and Pavlov, H. 2015, Eclipsing Binaries Add-in for TANGRA (http://www.hristopavlov.net/tangra3/eb.html).

Mousis, O., et al. 2014, *Experimental Astron.*, **38**, 91.

Nelson, R. 2014, Eclipsing Binary O–C Files (http://www.aavso.org/bob-nelsons-o-c-files).

Natural Satellites Data Center (NSDC). 2014, Light curve of J1 occulting J3 on May 8, 2009 (ftp://ftp.imcce.fr/pub/NSDC/jupiter/raw_data/).

Pojmański, G. 1997, *Acta Astron.*, **47**, 467.

Paschke, A. 2015, O–C Gateway (http://var.astro.cz/ocgate).

Pavlov, H. 2014, OCCUREC and TANGRA3—Software Packages for Astronomical Video Recording (http://www.hristopavlov.net).

Samolyk, G. 2015, private communication.

Sposetti, S., Lena, R., and Iten, M. 2012, *Lunar Planet. Inst. Contrib.*, No. 1659, 1012.

VideoTimers. 2011, IOTA-VTI (http://videotimers.com/).

Watson, C., Henden, A. A., and Price, C. A. 2014, AAVSO International Variable Star Index VSX (Watson+, 2006–2014; http://www.aavso.org/vsx).

Ye, Q.-Z. 2011, *Publ. Astron. Soc. Pacific*, **123**, 113.

Margaret Harwood and the Maria Mitchell Observatory

James W. Hanner
18 Alyssum Drive, Amherst, MA 01002; msh_jwh@yahoo.com

Received Jaunary 15, 2015; accepted May 21, 2015

Abstract The Maria Mitchell Observatory (MMO) and its first Director, Margaret Harwood, played an important role in encouraging women in astronomy.

1. Beginnings at Mt. Wilson Observatory

Mt. Wilson was the world's leading observatory from 1906 to 1949. During that period and beyond there was a strict prohibition against use of its facilities by women. The Director of the Observatory for much of the time was George Ellery Hale, who adhered strictly to established procedures. His health was intermittently precarious and in 1923 he took a leave of absence to recuperate. This circumstance allowed the one exception to the Mt. Wilson protocol: the acting director, Walter Adams, briefly relaxed the prohibition of women, to the benefit of Margaret Harwood (Figure 1).

Figure 1. Margaret Harwood 1886–1979. Director of Maria Mitchell Observatory 1912–1957.

Margaret Harwood spent most of her sabbatical, from August 1923 until June 1924, at Mt. Wilson. There she observed variable stars and asteroids, particularly Eros, a small asteroid with a close approach to Earth. She found, to the surprise of some astronomers, that its brightness varied (probably due to its rotation). During the last three months of her stay, she was granted the privilege of using the 60-inch telescope, formerly the largest in the world until supplanted by the 100-inch in 1916.

Harwood was friendly with Edwin Hubble, whose photo she took near the time of his discovery of Cepheid variables in the Andromeda Galaxy (Christianson 1995). Cepheids were key to supporting the argument that the spiral nebulae were galaxies far outside our Milky Way. Imagine an animated conversation between the two variable star astronomers, sharing the excitement of Hubble's discovery. Hubble and Harwood were partnered in one of two teams to observe the September 10, 1923, total solar eclipse. Their team, sent to Point Loma near San Diego, was frustrated to sit in their tents enjoying the rain. Miss Harwood had better luck with eclipses on Nantucket.

Prior to her stay at Mt. Wilson, Harwood had spent the spring of 1923 at the Harvard station at Arequipa, Peru. There, she studied the southern skies at altitude with the snow-capped volcano El Misti (19,000 feet) as a background. Her widely published photograph of the southern Milky Way was taken with a telescope of one inch in aperture.

2. Assistants at Maria Mitchell Observatory 1914–1956

Maria Mitchell Observatory (MMO) is a small facility on the island of Nantucket, Massachusetts, but it has been highly effective in encouraging women to work in astronomy. The tradition began with the first American woman astronomer, Maria Mitchell. Many women astronomers were invited to participate as assistants under the tutelage of Margaret Harwood, Director of MMO from 1912 to 1957 (Table 1). The term of residence varied from a summer to a year or more. MMO became a haven for young women—including the AAVSO's future Director, Margaret Walton Mayall (Figure 2)—to learn astronomical techniques and use of the telescope. These skills would be utilized during their careers of observing, research, publication, and teaching. The pattern altered in 1937, when Miss Harwood encouraged young men local to Nantucket to be assistants. The author participated in 1953, 1954, and 1955.

Figure 2. Margaret Harwood, student assistant Margaret Walton, and Harlow Shapley, advertising an "Open Night" talk by Shapley at the Maria Mitchell Observatory in 1925.

Five of the MMO assistants later received the Annie Jump Cannon Award, at that time presented by the American Astronomical Society (AAS) only once every three years to a distinguished woman astronomer.

Following Margaret Harwood's retirement the tradition was strengthened by astronomer Dorrit Hoffleit when she became MMO Director; her outstanding program (2 to 6 interns each summer) resulted in many who went on to distinguished careers in astronomy.

References

Christianson, G. E. 1995, *Edwin Hubble: Mariner of the Nebulae*, Farrar, Straus, Giroux, New York.

The Nantucket Maria Mitchell Association. Various issues, *Annual Report*, Nantucket Maria Mitchell Association, Nantucket, MA.

Table 1. Maria Mitchell Observatory assistants 1914–1956.

Year	Name	Affiliation while at MMO	Notes
1914	Susan Raymond	Harvard	
1918	Mary H. Vann	Harvard	
1919	Dorothy W. Block	Harvard	
1919	Mary D. Applegate	Vassar	
1920	Antonia C. Maury	Harvard	1, 4
1922	A. Grace Cook	England	
1924	Adelaide Ames	Vassar	
1924	Cecelia Payne	Harvard	2, 4
1925	Margaret L. Walton	Swarthmore	3, 4
1926	Margaret L. Walton	Swarthmore	3, 4
1929	Helen B. Sawyer	Mt. Holyoke	4
1930	Frances W. Wright	Harvard	
1931	Marjorie Williams	Smith	
1935	Helen W. Dodson	Wellesley	4
1937	Edgar Sanborn, Jr.	—	
1938	John H. Heath	—	
1943	Helen Wright	Vassar	
1948	Nan Reier	Goucher	
1949	Jocelyn Gill	Wellesley	
1950	Jocelyn Gill	Wellesley	
1951	Jocelyn Gill	Wellesley	
1953	James W. Hanner	—	
1954	James W. Hanner	—	
1955	James W. Hanner	—	
1956	John Turtle	—	

Notes: 1. Student of Maria Mitchell; 2. Otto Struve, Director of Yerkes Observatory and McDonald Observatory, AAS President (1946–1949) and IAU President (1952–1955), declared Payne's doctoral dissertation to be "undoubtedly the most brilliant Ph.D. thesis ever written in astronomy"; 3. Recorder, then Director of the American Association of Variable Star Observers (AAVSO) (1949–1973); 4. Recipient of Annie J. Cannon Award, American Astronomical Society (AAS).

Some Personalities from Variable Star History

Edited by

Thomas R. Williams
21 Briar Hollow Lane, Unit 406, Houston, TX 77027; trw@rice.edu

Michael Saladyga
AAVSO, 49 Bay State Road, Cambridge, MA 02145; mike@aavso.org

Received April 2, 2015; accepted May 29, 2015

Abstract Presented are sixty-six biographical sketches of noteworthy persons who have been associated with the AAVSO and variable star astronomy during its more than 100-year history.

1. Introduction

The original plan for *Advancing Variable Star Astronomy: The Centennial History of the American Association of Variable Star Observers* (Williams and Saladyga 2011) included sidebars comprising a selection of contributors to variable star astronomy. The list of proposed sidebars included both AAVSO members and noteworthy non-members. We solicited authors for each sidebar by publishing the proposed list of individuals. A gratifying response of volunteers to our appeal resulted in an interesting series of short articles. Regrettably, as the deadline approached to send the manuscript to Cambridge University Press we were unclear about meeting our contractually limited length of the book. To make certain there would not be an additional round of editing to come within that contractual limit, we reduced the manuscript's length by cutting out some important history, including the sidebars. By doing so, we insured that the published book would be available before the centennial celebration. To make sure the excised history, including the sidebars, would be published, this article is the first of others that will accomplish that goal. We are late in doing so, and must apologize to the authors of various sidebars for our tardiness. With our hearty thanks to all the authors—especially Tim Crawford—for their contributions, we hope you enjoy these fascinating vignettes.

2. The biographies

Note: The biographies are arranged chronologically according to each individual's early involvement with the AAVSO or variable star work. Published biographical or obituary articles, where available, are cited in the text and noted in the reference list. A name index is given in section 3.

Paul Sebastian Yendell (1844–1918)

by Matthew Templeton

Paul Yendell of Dorchester (Boston), Massachusetts, a private during the Civil War and grandson of a shipwright of *USS Constitution*, was a draftsman for the Massachusetts Topographical Survey and the Boston Public Works Department, and became one of the preeminent variable star astronomers of his time. He was a devoted, careful, and prolific visual observer for thirty years, making over 30,000 observations, even as many of his contemporaries lost interest and ceased observing. His observations and other writings frequently appeared in *Popular Astronomy* and *The Astronomical Journal*. Along with his meticulous observing skills, Yendell also possessed a New Englander's sensibility and dry humor: in one 1904 correspondence with Seth Chandler, Yendell noted: "Got an acre of stationery and 25 words of French politeness from [Michel] Luizet today, in acknowledgment of my U Cephei paper. [Joseph] Plassmann did the same thing with a postal card and about seven words of very compact German."

Edwin Forrest Sawyer (1849–1937)

by Michael Saladyga

One of the first amateur variable star observers in the U.S., Sawyer began observing as early as 1865. He joined the Boston Scientific Society in 1876 in the company of Chandler and Yendell. In 1893 Sawyer published a *Catalogue of the magnitudes of southern stars;* for which he made 13,654 observations to determine visual magnitudes of 1,145 stars, and in the process discovered eight variable stars and 51 suspected variables. Born in Boston and a bank cashier there for 64 years, he joined the AAVSO in 1921. (Taibi 2004)

Seth Carlo Chandler, Jr. (1846–1913)

by Matthew Templeton

S. C. Chandler began as an assistant to astronomer Benjamin Gould during the late 1860s but spent much of his professional life as an actuary while he was a devoted amateur astronomer. He briefly worked for the

Harvard College Observatory in the early 1880s, but conducted the majority of his astronomical work independently. Most noted for his discovery of the "Chandler wobble" (a minute, periodic perturbation of the Earth's rotational axis), he conducted extensive research on variable stars, published three variable star catalogues, and was editor of *The Astronomical Journal*. (Searle 1914)

Mary Watson Whitney (1847–1921)

by Kristine Larsen

Mary Whitney earned A.B. and A.M. degrees as astronomer Maria Mitchell's student and protégé, succeeding her as director of the Vassar College Observatory and Professor of Astronomy in 1888. Whitney established a student-based research program at Vassar, focusing on observations of comets, asteroids, and, after Nova Persei 1901, on variable stars. She is credited with the discovery of SX Draconis. In 1906 she developed an undergraduate course on variable stars, probably the first in the world, on which Caroline Furness based her 1915 textbook. During Whitney's tenure at Vassar, more than 100 articles and observing reports were published by Vassar staff and students. She and her students also sent variable star observation reports to Harvard College Observatory. (Furness 1922-1923)

Clara Hyde Olcott (1867–1951)

by Michael Saladyga

Clara Eunice Hyde was born in Norwich, Connecticut, where her family were among the founders of that town. She had a rigorous education at Norwich Free Academy and at Miss Dana's School, an elite finishing school in Morristown, New Jersey, but she did not attend college, choosing instead a vocation as music teacher. She met William Tyler Olcott in the 1890s; they married in 1902 and lived in the ancestral Tyler home in Norwich. During World War I she volunteered full time as chairwoman of the American Red Cross Volunteers. Described as "a commanding figure, a woman of great personal charm, and distinguished for many endearing traits of character," Clara Olcott encouraged and supported her husband in pursuit of his astronomical interests and in his dedication to the AAVSO. (Anon. 1951)

Anne Sewell Young (1871–1961)

by Michael Saladyga

As teacher and astronomer, Anne Young found the AAVSO to be the ideal training ground for her students and an important channel for conducting serious work in astronomy. Born in Bloomington, Wisconsin, her uncle was the well known astronomer Charles A. Young. She had one sister, Elizabeth, and the two of them remained close throughout their lifetime. Anne Young attended Carleton College in Northfield, Minnesota, receiving a Master's degree in 1897. She was principal of a high school in St. Charles, Illinois, but in 1899 found her life's work as head of the astronomy department and director of the John Payson Williston Observatory at Mount Holyoke College, South Hadley, Massachusetts. In 1902 She volunteered to observe variable stars for Harvard College Observatory, and in 1905 she enrolled at Columbia University, earning a Ph.D. there in 1906. She continued at Mt. Holyoke while volunteering as a research assistant at Yerkes Observatory during the summer months. Anne Young was one of the seven original members of the AAVSO, and she encouraged her students to attend AAVSO meetings. Three of her students—all AAVSO members—became well known astronomers: Louise Jenkins, Alice Farnsworth, and Helen Sawyer Hogg—and one of Farnsworth's students, also at Mt. Holyoke, was Martha Hazen. Young, Farnsworth, Hogg, and Hazen were each influential presidents of the AAVSO. Anne Young retired in 1936 and lived with her sister in Claremont, California. (Bracher 2012)

Charles Y. McAteer (1865–1924)

by Michael Saladyga

Of the early AAVSO members known as "The Old Guard," none was better liked and more respected than C. Y. McAteer. "Mac" lived in Pittsburgh, Pennsylvania, where he was an engineman for the Pittsburgh, Cincinnati, Chicago & St. Louis Railway Co. From 1912 to 1924 he made over 12,200 variable star observations. The earliest were made with a 4-inch refractor which he would set up in his yard in the pre-dawn hours after coming off of his all-night fast freight runs. Self-taught in astronomy, McAteer found a kindred spirit in the celebrated Pittsburgh telescope maker John Alfred Brashear, who had spent twenty years laboring as a rolling mill mechanic. The two men became close friends. McAteer's observing skill and knowledge of astronomy was noted by the staff at the University of Pittsburgh's newly constructed Allegheny Observatory, who allowed him to use its 13-inch refractor to make his variable star observations; he was also permitted to run the public observing sessions there. McAteer died tragically in 1924, struck by a truck as he ran for a trolley car on his way home from work. Soon afterwards, the

AAVSO library—which McAteer had begun in 1921 with a number of books bequeathed to him by Brashear—was officially named the "Charles Y. McAteer Library of the AAVSO." (Anon. 1924)

Giovanni Battista Lacchini (1884–1967)

by Michael Saladyga

G. B. Lacchini was with the AAVSO for over 50 years, beginning with his first contact with W. T. Olcott in 1911 as its first international member. He made over 65,000 variable star observations from 1911 to 1963 and served on the council (1930–1932; 1961–1963). Lacchini was born in Faenza, Italy, where he had his observatory at his childhood home. An amateur turned professional, Lacchini was a postal worker until 1928 when he was named Assistant Astronomer for the Italian Government at the Royal Observatory in Catania, Sicily. Later he worked at Torino Observatory, and then at Trieste Observatory. His Faenza home and observatory were completely destroyed in bombings during World War II. (Anon. 1967)

Edward Gray (1851–1920)

by Michael Saladyga

Physician Edward Gray, born in Benicia, California, one of the first seven AAVSO members, was also head of the Variable Star Section of the Society for Practical Astronomy. Gray understood the scientific value of variable star work and soon threw his full support over to the AAVSO. He introduced the blueprint method of reproducing variable star charts which was then refined by AAVSOer Harry C. Bancroft, Jr. Gray also wrote fiction and text books, and was considered an authority on the Spanish language. For the last four years of his life he was ship surgeon for the Pacific Mail Steamship Company, assigned to a San Francisco-New Zealand route. (Mundt 1920)

Tilton Clark Hall Bouton (1856–1948)

by Tim Crawford

An ordained minister, T. C. H. Bouton joined the AAVSO in 1911. He was also member of the Society for Practical Astronomy, Variable Star Section. In 1912 Bouton wrote in *Popular Astronomy* of a box lantern device to hold charts, with needle holes for the stars, so that the eyes could remain dark adapted for variable star work. He reported 24,884 observations to the AAVSO between 1914 and 1947 and received the second AAVSO Merit Award in 1936.

David Bedell Pickering (1873–1946)

by Tim Crawford

David Pickering, a jeweler, joined the AAVSO in 1914 and attended the Annual Meeting held in Cambridge, Massachusetts, in 1915. He hosted, at his home in East Orange, New Jersey, the first AAVSO Spring Meeting in 1916 and continued hosting spring meetings through 1919. Pickering was the first president of the association (1917–1918), a Patron (1919), and the sixth president (1927–1928). He chaired its first chart committee, formed circa 1918–1919, and held that position until 1934. Pickering received the fifth AAVSO Merit Award in 1938 "in appreciation of his long leadership in developing the Association and standardizing its methods...." (Campbell 1946)

Caroline Ellen Furness (1869–1936)

by Kristine Larsen

After completing her A.B. and A.M. degrees as a student of atronomer Mary Whitney at Vassar College, Caroline Furness became the first woman to earn a Ph.D. in astronomy at Columbia (1900). She collaborated with Whitney as her assistant between 1909 and 1911, each sending their variable star observations to Harvard College Observatory. A member of the AAVSO from 1911, Furness succeeded Whitney in 1913, and prepared for publication a volume of variable star observations made at Vassar from 1901 to 1912. In 1915 she authored the well-received *An Introduction to the Study of Variable Stars*. Furness involved her students in variable star observing and AAVSO Annual Meetings. (Makemson 1936)

Helen Margaret Swartz (1878–1959)

by Dee Sharples

Helen Swartz, born in Norwalk, Connecticut, and a member of the AAVSO from 1911, was elected to its first council in 1917. As a student of astronomer Mary Whitney at Vassar College, she earned B.A. (1901) and M.A. (1903) degrees. For several years, she taught astronomy, mathematics, and German in the Norwalk High School and served as one of Edward Pickering's Harvard Observatory corps of observers. Beginning in October 1911 she sent her observations to the AAVSO. Using a 3-inch refractor, she made the first observation of R Cyg to be archived in the

AAVSO International Database. She was one of the founders of the Norwalk Astronomical Society and a member of the American Astronomical Society. (Anon. 1959)

Willem Jacob Luyten (1899–1994)

by David B. Williams

Dutch-American astronomer W. J. Luyten was the last living charter member of the AAVSO and a speaker at the 1986 75th anniversary banquet and headquarters building dedication. He made visual observations of variable stars as a teenager in Holland. At Harvard and then the University of Minnesota, he discovered thousands of high proper-motion stars in a search for nearby white dwarfs. As a bonus, he discovered more than 1,700 new variables during the Bruce Proper Motion Survey of the southern sky. He detected the first flare of his PM star L 726-8, now known as the prototype flare star UV Ceti. (Luyten 1995)

Charles Francis Richter (1900–1985)

by Kate Hutton

As one of the AAVSO's prolific early observers and a charter member, Los Angeles teenager Charles Richter (RCF) submitted 7,190 variable star estimates between 1914 and 1917. Later in life, he put his familiarity with magnitudes to good use, devising the first logarithmic method of measuring the sizes of earthquakes. Although he called his system "local magnitude," since it was used for earthquakes relatively near to the seismographic stations, the world of popular science universally knows it and its successors as "the Richter scale." (Wilford 1985)

J. Ernest G. Yalden (1870–1937)

by Michael Saladyga

Known to early AAVSOers as "The Baron," James Ernest Grant Yalden's presence at AAVSO meetings set a tone of good cheer that long outlived him. Born in Brixton (London), he came to the U.S. with his family when he was about ten years old. While a student at New York University he befriended the artist Winslow Homer. The two spent summers camping in the Adirondacks; Homer's watercolor "Paddling at Dusk" depicts Yalden in his handmade canoe. He married Margaret Stewart Lyon in 1895 in New York City where Yalden became Superintendent of the Baron de Hirsch Trade School. The Yaldens moved to Leonia, New Jersey, in 1901. Yalden joined the AAVSO in 1918 and became active on committees for charts, telescopes, and occultations. He was named a Patron of the association in 1921, served as vice-president in 1925, and president in 1926. (Farwell 1937; Ingham 1937; Pickering 1937)

Radha Gobinda Chandra (1881–1975)

by Tim Crawford

After observing Comet Halley 1910, Chandra, a civil servant in Bagchar, Jessore, India, purchased a small telescope. A self educated astronomer, he reported his discovery of Nova Aquilae No. 3 to Harvard Observatory in 1918, and then joined the AAVSO. The association loaned him a 6–inch telescope in 1926, donated by Charles W. Elmer. Chandra was a member of the British Astronomical Association-Variable Star Section and Association Française des Observateurs d'Étoiles Variables, in 1928 he received the French Officer d'Academic Brevet and Badge for his astronomical work. From 1920 to 1954 Chandra reported a total of 37,215 observations. The Elmer telescope passed to M. K. Vainu Bappu, who earned a Ph.D. in astronomy at Harvard. Chandra helped found astronomy education programs for at least two institutions for young people in India. (Bandyopadhyay and Chakrabarti 1991)

Leslie Copus Peltier (1900–1980)

by Tim Crawford

In 1917 Leslie Peltier picked strawberries to purchase his first telescope, a 2-inch refractor. That Christmas his mother gave him a copy of William Tyler Olcott's *A Field Book of Stars*. Olcott encouraged readers to contact him regarding how they could engage in astrophysical research. Peltier did so, and became a lifelong member of the AAVSO in 1918. He made regular monthly contributions, reporting over 132,000 observations until his death. He also discovered eleven comets and two novae, authored three books, wrote numerous articles, and received the AAVSO's first Merit Award in 1934. Harlow Shapley described him as the "world's greatest non-professional astronomer." (Hurless 1980a, 1980b)

Charles Cartlidge Godfrey (1855–1927)

by Michael Saladyga

Charles C. Godfrey was born in Saybrook, Connecticut, and resided in Bridgeport. Onetime personal physician to showman P. T. Barnum, and his state's surgeon general, Godfrey also helped found the Bridgeport Library, organized the city's school system, improved the city water system to reduce transmission of disease, and helped organize the city's scientific and historical society. He was keenly interested in all science and technology of his time, especially electricity and radio. Described has having a "firm, cheerful, and courageous character," Godfrey joined the AAVSO in 1919, was a council member (1925–1927), president (1926–1927), and a Patron of the association (1927). He became fast friends with the other early AAVSOers, camping in Maine with J. E. G. Yalden and Leon Campbell. As a wry expression of friendship, surgeon Godfrey once said to Yalden "come and visit me in Bridgeport and I'll cut your leg off for you." (Waldo 1928)

Ralph Noyes Buckstaff (1887–1980)

by Michael Saladyga

Ralph Buckstaff was born in Oshkosh, Wisconsin. He worked for The Buckstaff Company of Oshkosh, a large furniture manufacturer started by his grandfather, and became its president. Buckstaff belonged to many scientific societies, was a founding member of the Milwaukee Astronomical Society, and joined the AAVSO in 1920. A lifelong interest was meteorology, and he served for 56 years as a volunteer regional weather observer for the U.S. Department of Commerce. His passion, however, was astronomy, and he was especially interested in cooperation with schools and other groups to make astronomy interesting and available to all. The AAVSO awarded him its 18th Merit Award in 1965 in recognition of his efforts in astronomy outreach. He was acknowledged to be "the epitome of a scientific citizen and worked to communicate his interests to students of all ages—giving generously of his time and talents." One of those young astronomers he sought out and offered to help was Walter Scott Houston, who later wrote that, of all the awards and honors bestowed by many organizations, Ralph Buckstaff considered his AAVSO Merit Award "his most valued recognition." (Ekvall 1979-1980)

Charles W. Elmer (1872–1954)

by Tim Crawford

Charles Elmer joined the AAVSO in 1920, was named a Patron of the association in 1921, and served as president for the 1937–1938 term. He received the seventh AAVSO Merit award in 1943. In 1927 Elmer was a principal in establishing the popular Custer Institute and Observatory on Long Island, New York. Elmer was actually nearing retirement as head of a firm of court reporters when he joined with Richard Perkin as co-founder of the Perkin-Elmer Corporation in 1937. Perkin shared Elmer's astronomical interest, but whereas Elmer preferred to spend his time in their workshop, Perkin ran the corporation. During WWII Perkin-Elmer produced optics for bombsights and rangefinders. (Mayall 1955)

Dalmiro Francis Brocchi (1871–1955)

by Thomas R. Williams

Through the middle years of the 20th century, visual observers of variable stars used AAVSO blueprint charts drawn by an Italian-American railroad draftsman, D. F. Brocchi. These meticulous charts featured small-diameter stars with fine variations in star diameter to reflect differences in magnitude. He prepared these charts by tracing star fields from photographic plates taken at various observatories and from star field sources such as the *Bonner Durchmusterung*. Brocchi also observed variable stars in two observatories he built at his home in Seattle, Washington. His largest telescope, a 12-inch spherical mirror coupled with a Harris prism, was mounted on a unique hemispherical axis with the sky engraved on the surface of the hemisphere. A pointer shadow on the hemispherical map indicated the position towards which the Equatazimuth Telescope (a term he coined) was pointing. Brocchi joined the AAVSO in 1921, was chart committee chair 1934–1948, served on council 1936–1938 and 1940–1941, and was named a Patron of the association in 1947. In 1942 Brocchi received the sixth AAVSO Merit Award.

Percy W. Witherell (1877–1970)

by Tim Crawford

Percy Witherell, an MIT graduate and treasurer of a family-owned grocery business, joined the AAVSO in 1921, serving as treasurer from 1931 to 1960. He received the 12th AAVSO Merit Award in 1953 in appreciation of his loyal and devoted service. During the early 1940s he authored thirty-six articles and reports and gave other support to the fledgling *Sky & Telescope* Magazine. Witherell also served as president of the Bond Astronomical Club in Cambridge, Massachusetts, which later merged into the Amateur Telescope Makers of Boston. (Anon. 1970a)

Alice Hall Farnsworth (1893–1960)

by Thomas R. Williams

As an undergraduate at Mount Holyoke College, Alice Farnsworth studied astronomy for four years with Anne Sewell Young, receiving an A.B. degree in 1916. At the University of Chicago and Yerkes Observatory, she learned photometry from noted variable star astronomer John Adelbert Parkhurst. With her Ph.D. in hand, in 1920 Farnsworth returned to Mount Holyoke as an instructor working with Anne Young until the latter retired in 1936. Farnsworth joined the AAVSO in 1921 and spent the remainder of her career as chairman of the Mount Holyoke Astronomy Department and director of the Williston Observatory. She retired in 1956 after suffering a stroke. In addition to timing lunar occultations, Farnsworth assisted the AAVSO occultation committee with reduction of occultation observations. Farnsworth also observed sunspots daily for the AAVSO and served on the AAVSO council for ten years including her term as president (1929–1931). (Anon. 1960a, 1960b)

Lewis Judson Boss
(1898–1982)

by Tim Crawford

Boss, an electrical engineer, joined the AAVSO in 1921. A pioneer in photoelectric photometry using selenium cells in self-made photometers, he presented a paper at the AAVSO's Annual Meeting in 1922 describing his early results measuring variable stars with his equipment, which he continued to improve. In 1954 he became the first chairman of the AAVSO photoelectric photometry committee, holding this position until 1967. Boss served on the AAVSO council 1943–1947 and 1962–1964. Along with variable stars his interests included eclipses, transits, and aurorae. He authored many articles and papers.

Morgan Cilley
(1878–1955)

by Tim Crawford

Cilley, a civil engineer, a U.S. Naval Observatory Staff member in later years, and an ordained Episcopal minister in between, joined the AAVSO in 1923. He was considered one of the AAVSO's "Old Guard" members and served on the council from 1942 to 1944. He made 15,804 observations for the AAVSO between 1923 and 1950. Cilley authored several articles for *Popular Astronomy*. As a civil engineer, he worked on projects including the building of the Pennsylvania Railroad tunnels under the Hudson River (circa 1902) and on road projects in the Philippine Islands (circa 1909). He retired in Lewisburg, West Virginia. (Anon. 1955)

Eugene H. Jones
(1864–1946)

by Michael Saladyga

"Jonesey" as he was popularly known by his fellow AAVSO members, made 44,764 variable star observations between 1924 and 1944. He joined the AAVSO in 1923 at age 59 and served on the council from 1933 to 1936. Eugene Jones was born in Cambridge, Massachusetts, and from age 10 lived just north of there in Somerville. He began working for the Boston Edison Company in Somerville in 1888 as an inspector of electric street lamps, and continued at that work until he retired in 1929. He then moved to Goffstown, New Hampshire, where he built an observatory and made observations until his health failed. He was presented the AAVSO's fourth Merit Award in 1937. His brother Fred, also an AAVSO member, was known for designing and hand-illuminating the first thirteen AAVSO Merit Award scrolls. In addition to being an excellent and dedicated observer, Eugene Jones was an amateur landscape painter, photographer, and musician. One of his paintings can be seen in a photo of Leon Campbell's office in 1936; it is a view of his Goffstown observatory in winter. (Rosebrugh 1937)

Helen Battles Sawyer-Hogg-Priestley
(1905–1993)

by Thomas R. Williams

Born in Lowell, Massachusetts, Helen Sawyer graduated from Mount Holyoke College. She met and married Canadian Frank Hogg while a graduate student at Radcliffe College and Harvard College Observatory. Cecelia Payne-Gaposchkin and Harlow Shapley guided Sawyer-Hogg's thesis work. She accompanied Frank to Canada in 1931 and completed her career of 35 years at the University of Toronto. Recognized as the international authority on variable stars in star clusters, her three catalogues of these objects remained standard resources for many years. In addition to her observational work and teaching, she established herself as the leading popularizer of astronomy in Canada through her regular columns in the Toronto *Daily Star*, on television programs, and as the author of *The Stars Belong to Everyone* (1976). An active supporter of the AAVSO, she served as chart curator, and served fourteen years on the council, including two terms as president. She married professor Francis Priestley in 1985. (Cahill 2012; Clement 1993; Broughton 1994)

Harold B. Webb
(1896–1976)

by Thomas R. Williams

Webb joined the AAVSO in 1928 and contributed 9,213 observations between 1928 and 1953. As a professional draftsman, it was natural for Webb to serve on the AAVSO chart committee. He was named curator of tracings and printing in 1943. Webb established a printing business in the mid-1940s, and by 1954 his published titles included three books on Mars and four star atlases. Of the latter, the most popular were *The Observer's Star Atlas* and *Webb's Atlas of the Stars*. A limited edition of the *Atlas of the Stars* included long period variable stars. Although drafted with a limiting magnitude of about 9, *Atlas of the Stars* never gained popularity among AAVSO observers.

Helen Meriwether Lewis Thomas (1905–1997)

by Michael Saladyga

Helen Thomas joined the AAVSO in 1929 and was Pickering Memorial assistant to Leon Campbell at Harvard College Observatory from 1934 to 1937. During that time she assisted with HCO's Milton Bureau variable star survey, making 50,000 photographic magnitude estimates for over 100 stars, and, as she was proud of saying, she re-catalogued the AAVSO library. Thomas was the third person to earn a Ph.D. in the History of Science at Harvard or Radcliffe Colleges, and the first American woman to do so. Her 1948 dissertation is titled "The early history of variable star observing to the XIX century"; Dorrit Hoffleit declared this work to be "a true masterpiece." Thomas also assisted Willem Luyten with making proper motion measurements in 1928, is credited with the discovery of a white dwarf star, and rediscovered Nova Sco 1863 (U Sco), and, through her research, confirmed the star as a recurrent nova. In 1947 she began work as an electronics engineer, and in 1954 she was editor for the MIT Laboratory of Electronics, eventually becoming head of its Publication Office before retiring in 1971. Helen Thomas attended the AAVSO's 75th anniversary meeting in 1986. (Hoffleit 2000)

Winifred Crosland Kearons (1883–1958)

by Michael Saladyga

Winifred Kearons distinguished herself in her time by becoming the only woman in the AAVSO who, at her peak, made over 1,000 variable star observations each year. She amassed a lifetime total of 10,010 observations over 22 years and was one of 25 observers on Leon Campbell's "Roll of Honor" in 1946. Born in Rochester, New York, she and her husband, Rev. William M. Kearons, resided in Fall River and West Bridgewater, Massachusetts, where Rev. Kearons was an Episcopal minister. Both were experienced solar observers before they joined the AAVSO, providing counts to the sunspot bureau at Berne, Switzerland. Winifred Kearons joined the AAVSO in 1925, served on the council 1939–1943, and was named a Patron of the association in 1953.

Phoebe Waterman Haas (1882–1967)

by Dee Sharples

Phoebe Waterman Haas received a master's degree from Vassar College in 1906 and became a computer at Mount Wilson Observatory. She was the first woman to complete work on a Ph.D. in Astronomy (1913) at the University of California-Berkeley/Lick Observatory. Haas submitted 338 observations to the AAVSO between 1928 and 1933. More importantly, when Harvard curtailed its financial support of AAVSO in 1953, Haas provided vital support to Margaret Mayall by calculating the five- or ten-day means for southern variable stars. Continuing this work for more than ten years, her data formed the basis for light curves published by the AAVSO. (Williams 1991)

Leo John Scanlon (1903–1999)

by Thomas R. Williams

Plumber Leo Scanlon of Pittsburgh, Pennsylvania, built his first telescope in 1928. He co-founded the Amateur Astronomers Association of Pittsburgh (AAAP) in 1929, then joined the AAVSO in 1930. He gained national prominence in telescope making in the mid-1930s. After his marriage in 1940, he scaled back his interests to telescope making and popularizing astronomy as a planetarium lecturer/demonstrator. Widely known for having built Valley View, the first observatory with an aluminum dome, Scanlon's main contribution to variable star observing may have been his attempt to promote efficiency in observing and encouragement of other observers as an AAVSO regional advisor. (Callum 1937)

Ellen Dorrit Hoffleit (1907–2007)

by Kristine Larsen

After receiving her B.A. at Radcliffe College, Dorrit Hoffleit began a distinguished research career at the Harvard College Observatory (1929–1956) during which she earned an M.A. and Ph.D. from Radcliffe. She joined the AAVSO in 1930. From 1943 to 1948 she worked as a ballistics mathematician for the Aberdeen Proving Grounds. In response to changes in the HCO leadership, Hoffleit moved to Yale University, where she was an astronomy researcher from 1956 to 1983. After her formal retirement, she continued to work as a consultant and volunteer until shortly before her death. Hoffleit's research interests were myriad, ranging from stellar spectra, meteors, and variable stars to the history of astronomy. She authored several editions of the seminal *Bright Star Catalogue*. Perhaps her greatest contribution to astronomy remains her development of a summer research program for undergraduates while director of the Nantucket Maria Mitchell Observatory (1957–1978). Many of its alumnae have become professional astronomers and astronomy educators. Hoffleit was one of the AAVSO's most enthusiastic supporters and meeting attendees; she served on the council for a total of 23 years (including two as president). She received the third AAVSO William Tyler Olcott Distinguished Service Award (Hoffleit 2002; Anon. 2007; Larsen 2012a)

Walter Scott Houston
(1912–1993)

by Glenn Chaple

For nearly a half century Walter Scott Houston wrote the "Deep Sky Wonders" column for *Sky & Telescope*, but variable stars and the AAVSO were his first love in astronomy. In 1931, while Houston was a student at the University of Wisconsin, his friend Joseph Meek persuaded him to join the AAVSO. During the next six decades he served six terms on the council and contributed over 12,000 visual observations. In 1958, he directed the Manhattan, Kansas, Project Moonwatch Station that made the world's first sighting of Explorer I—the first American satellite. The minor planet 3031 *Houston* was named in his honor. (Anon. 1994a, 1994b)

David W. Rosebrugh
(1899–1988)

by Tim Crawford

Rosebrugh, an electrical engineer, joined the AAVSO in 1932. Born in Canada and a life-long member of the Royal Astronomical Society of Canada (RASC) he lived most of his life in the U.S. He served multiple terms on the AAVSO council, served as secretary from 1937 to 1945 and president for the 1948–1949 term. Rosebrugh was both an active visual variable star observer and sunspot observer, receiving the AAVSO's 11th Merit Award in 1951. An author with several papers and numerous articles in both *JRASC* and *Sky & Telescope*, Rosebrugh wrote in later years of observing Halley's Comet and Comet 1910a when he was eleven years old. (Anon. 1988; Broughton 1994)

Neal J. Heines
(1892–1955)

by Michael Saladyga

Born in Rotterdam, Neal Heines came to the U.S. with his parents when he was one year old; they settled in Patterson, New Jersey. He was part of a U.S. Army entertainment troupe in Europe during World War I, and for most of his life he was a commercial-industrial salesman. When he was in his thirties, he discovered that his great-great-grandfather was Eise Eisinge of Franeker, West Friesland, who constructed a large, ceiling-mounted orrery; this inspired Heines to learn all he could about astronomy. J. E. G. Yalden introduced him to the AAVSO in 1934. In 1944 Heines became the first chairman of the Solar Division; he established the *Solar Bulletin* in 1945. Heines was president 1949–1951, during which time he was co-author, with Harlow Shapley, of a "good will" letter to foreign astronomers. He was named a Patron of the association in 1946. Besides variable star and solar astronomy, Heines' scientific interests included general astronomy, seismology, and meteorology. His other interests included poetry, fly-fishing, gardening, arts and crafts, and music, and he conducted two glee clubs, a choral group, and a church choir. Those who knew him considered Heines to be a modern-day Renaissance Man. (Bondy 1955)

Reginald Purdon de Kock
(1902–1980)

by Thomas R. Williams

The AAVSO's leading observer at the time of its 75th anniversary, Reginald de Kock achieved that distinction in spite of significant disabilities that might have limited less-determined individuals. He never let the restricted use of his left arm nor the limited peripheral vision in his left eye inhibit his observation of southern hemisphere variable stars. As a daytime computer at the Royal Observatory at the Cape of Good Hope, South Africa, de Kock earned the right to estimate variable stars with observatory telescopes, and then began sending his observations to the AAVSO in 1934. A gentle, reserved person with a sly sense of humor, de Kock was regarded with affectionate respect by the observatory staff. The Royal Astronomical Society presented their Jackson-Gwilt Medal and Gift to de Kock in 1957, while the AAVSO elected him to honorary membership in 1946, and awarded him its 15th Merit Award in 1961 to recognize his "constant vigilance of the morning and evening sky...." (Glass 1986)

Edward Anthony Halbach
(1909–2011)

by Robert Stencel

Ed Halbach submitted more than 98,000 variable star observations to the AAVSO. A co-founder of the Milwaukee Astronomical Society in 1932, an AAVSO member from 1934, and well known in amateur astronomy, he served as first president of the Astronomical League in 1947. Later, he was active observing lunar occultations, orbiting satellites (Project Moonwatch), eclipsing binaries and other variable stars. For leading six international solar eclipse expeditions for the federal government and National Geographic Society, the latter honored him with its Franklin L. Burr Award. Halbach received the 32nd AAVSO Merit Award in addition to its fourth William Tyler Olcott Distinguished Service Award and the Astronomical League's Leslie C. Peltier Award. (Williams 2007; Samolyk 2011)

Richard Warren Hamilton
(1918–1975)

by Michael Saladyga

Richard Hamilton, born in New York, was a great-great-grandson of the first U.S. Treasury Secretary, Alexander Hamilton. His mother, Mary Clark Spurr Hamilton, descended from settlers of the Plymouth colony in Massachusetts. The Hamiltons lived in Norwalk, Connecticut. There he met AAVSO charter member Helen Swartz, who sponsored the sixteen-year-old Hamilton for AAVSO membership in 1934—she and his mother were both members of the Norwalk Astronomical Society. His mother, also an AAVSO member, would attend AAVSO meetings with him until her death in 1952. Richard Hamilton served on the council 1949–1960 and 1968–1970. He was president 1956–1958, chart curator 1948–1961, and chart committee chair 1955–1969. During these years he lived in Fairfield and Darien, Connecticut.

J. Russell Smith
(1908–1997)

by Michael Saladyga

Joseph Russell Smith, born near Walnut Springs, Texas, joined the AAVSO in 1936. He made 3,346 observations of faint variable stars in his 44 years with the AAVSO. He can, perhaps, be considered a typical member who sent observation reports at every opportunity and was content with knowing that his work was contributing to knowledge about the stars. Smith was a high school science teacher all his life, which was interrupted only by his service in the U.S. Army in Europe during World War II which included being part of the Normandy Invasion. He returned from the war, resumed his teaching and observing, and published a well-received book on teaching astronomy. He made his pre-war observations with an 8-inch reflector in Smyer, Texas, and his later observations with a 16-inch reflector in Eagle Pass, and then in Waco. He was one of the five founders of the Association of Lunar and Planetary Observers (ALPO) and served as its secretary. Smith observed until old age forced him to give up his observatory; in 1991 Preston F. Gott acquired the observatory and moved it to Texas Tech University in Lubbock. (Smith 1938, 1959, 1973)

Francis Lancaster Hiett
(1915–2004)

by Tim Crawford

As a protégé of Morgan Cilley, Lancaster Hiett joined the AAVSO in 1936. With strong support from his wife Bernice, variable star observing became an important part of his life. He contributed over 115,000 visual observations in over sixty-two years of observing, and submitted important observations to the *IAU Circulars*. Hiett received the AAVSO's 27th Merit Award in 1986. He is credited with an independent discovery of C/1948 V1, named "Eclipse Comet of 1948." A junior high school science teacher, Hiett sponsored a junior astronomy club in Virginia, and was a member of the National Capital Astronomers in Washington, D.C.

Cyrus F. Fernald
(1901–1979)

by Tim Crawford

Cyrus F. Fernald, a Certified Public Accountant, joined the AAVSO in 1937. He served as president for the 1954–1955 term. In 1947 Fernald received the AAVSO's ninth Merit Award. In 1986, during the AAVSO's 75th Anniversary, Fernald was recognized as one of the three top observers with 134,582 observations. It was estimated that he averaged twenty observations per hour. He also became a sunspot observer in 1945 and with deteriorating vision in later years he became increasingly involved with sunspot counting. Fernald's wife, Emily, also was an observer and they left a substantial bequest to the AAVSO. (Peltier 1979; Anon. 1980)

Claude B. Carpenter
(1902–1992)

by Michael Saladyga

Claude Carpenter joined the AAVSO in 1937, and served as council member (1939–1943). A post office dispatcher and former electrician from Wayne, Michigan, he was president of the Detroit Astronomical Society. He supervised construction of, and assembled, a 16-inch reflector for J. Russell Smith. He retired to California in 1952 and eventually agreed to move his own 18-inch reflector to the Ford Observatory in Wrightwood, California, where he, Clint Ford, Ron Royer, Tom Cragg, and others put it to good use. Carpenter made his last observations at the Ford Observatory in 1978. (Anon. 1993)

Martha E. Stahr Carpenter
(1920–2013)

by Kristine Larsen

Martha "Patty" Carpenter is the only person to have served three consecutive terms as president of the AAVSO. Born in Bethlehem, Pennsylvania, she received a B.A. from Wellesley, and M.S. and Ph.D. degrees in astronomy from the University of California-Berkeley. She joined the AAVSO in 1939, and became a radio astronomer at Cornell University in 1947 and was the first woman faculty member in the College of Arts and Sciences. There she studied radio emissions from the Sun and produced the first bibliographic compendiums of worldwide radio astronomy research. She collaborated with J. Kerr and J. V. Hindman on the use of 21-cm radio observations to extend the map of the spiral structure of the Milky Way to the Southern Hemisphere. The AAVSO elected her to the council in 1946 and she became president in 1951. During her second term as president, the AAVSO was forced to become independent from Harvard and the council made the historic decision to elect her to an unprecedented third consecutive term. Carpenter afterwards served on the Endowment Committee and again on the council (1970–1973). She joined the faculty of the University of Virginia in 1969. While there she encouraged then graduate student Janet Akyüz (later Mattei) to become involved with the AAVSO, and was a vocal supporter of Janet's application to become Director of the organization. (Larsen 2012b)

Walter P. Reeves
(1884–1957)

by Michael Saladyga

Vice-president of the Maine Central Railroad, Walter Reeves joined the AAVSO in 1941 at age 57. He was born in Portland, Maine, and as a teenager began at MCR as a freight auditor's clerk, working his way up to auditor, comptroller, and vice-president. Reeves made only 198 variable star observations for the AAVSO, but he and his wife, Gertrude E. Morse Reeves, were strong supporters of the association and attended AAVSO meetings from as early as 1936. Walter Reeves served on the council (1953–1957) and the finance committee (1954–1957) at a time of great financial difficulty for the association; he was also a member of the Astronomical Society of Maine. Mrs. Reeves was named a Patron of the association in 1958.

Frank M. Bateson
(1909–2007)

by Elizabeth O. Waagen

Frank Bateson organized variable star observing in New Zealand, providing leadership to the field in the Southern Hemisphere for 78 years. In 1927 he founded the Variable Star Section (VSS) of the Royal Astronomical Society of New Zealand (RASNZ) and remained as director of the VSS until 2004. Born in Wellington, New Zealand, Frank Bateson worked as an accountant during his early years. He served in the Royal New Zealand Navy during World War II, after which he became director of a trading company. He joined the AAVSO in 1944. In 1960, he conducted site surveys for a proposed observatory to be jointly operated by the Universities of Pennsylvania and Canterbury. He and his wife, Doris, formed a non-profit organization called Astronomical Research Ltd. which administered the over one million observations which had been delivered to the VSS since the start of the program. Maintaining a close working relationship with the AAVSO, one of his most valuable contributions to the organization was his willingness to share information on countless numbers of Southern Hemisphere variables. The AAVSO awarded him an Honorary Membership in 1986. (Anon. 2007; Christie 2007)

Helen M. Stephansky Abbott
(1919–   )

by Michael Saladyga

Helen Stephansky began working for Leon Campbell in 1944 as the Harvard College Observatory Pickering Memorial assistant. She continued at that job after Campbell retired in 1949 when Margaret Mayall took over as AAVSO director. Mayall and Stephansky worked well together and became good friends, having a common bond of optimism, confidence, and good sense. Although not an amateur astronomer herself, she appreciated what AAVSOers accomplished, was supportive of their efforts, and felt genuine affection for members she came to know. During the AAVSO's years of crisis in the 1950s, Stephansky proved herself a capable and loyal worker. By 1960 the AAVSO was on firmer footing, and Stephansky resigned to seek work elsewhere in the Boston area. In gratitude for her loyalty and service to the AAVSO, the council elected her to Life Membership. She married Herbert Whipple Abbott in 1961 and they eventually settled in northern New Hampshire.

Albert F. A. L. Jones
(1920–2013)

by Elizabeth O. Waagen

Albert Jones of Nelson, New Zealand was the world's most prolific visual variable star observer, with over 500,000 observations made between 1943 and 2011. Blessed with excellent eyesight and powers of discrimination, his location in New Zealand made him an even more crucial observer because of the relatively small number of variable star observers in the Southern Hemisphere. He followed many of his stars for decades, creating priceless, uninterrupted light curves that in numerous cases contain all or much of the optical data in existence for those stars over those intervals. He was interested in astronomy from boyhood and it became a fundamental part of him for the rest of his life, even as he earned his living as a miller in an oat cereal mill (as had his father), a grocery shop owner, and a worker in a car assembly factory. Albert Jones began observing variable stars at about age 23, and joined the AAVSO in 1997. In addition to AAVSO Observer Awards, Albert Jones received the AAVSO Director's Award in 1997, and the 41st AAVSO Merit Award in 2008. The AAVSO made him an Honorary Member in 2011. (Anon. 2013b)

Thomas A. Cragg
(1927–2011)

by Michael Saladyga

When he joined the AAVSO in 1945 at age 17, Tom Cragg was already working as an assistant for the Mt. Wilson observatory; his career as a professional/amateur astronomer led to 157,056 variable star observations for the AAVSO, and an almost uninterrupted series of daily sunspot observations and drawings. Born in St. Louis, Missouri, Cragg lived in Los Angeles until about age 48 when he relocated to Australia where he worked at the Siding-Spring Observatory. While in California, he was one of the group of observers who used the Ford Observatory, including Clint Ford, Ron Royer, and Claude Carpenter. One of Cragg's most important contributions was in chart and sequence work and expanding the observing program. He served on the council (1951–1953, 1962–1966) and chaired the classical cepheids committee (1967–1994). He was presented with the AAVSO's 25th Merit Award (1986). (Toone 2011)

Lawrence N. Upjohn
(1873–1967)

by Michael Saladyga

With experience in astronomy that was more casual than scientific, Lawrence Northcote Upjohn, M.D., nevertheless helped to advance variable star astronomy as a generous benefactor. Upjohn, the nephew of the founder of pharmaceuticals manufacturer The Upjohn Company, in Kalamazoo, Michigan, was its chief executive from 1932 to 1934, and from 1944 was its chairman of the board. Upjohn's interest in astronomy started at boyhood, no doubt influenced by his grandfather Uriah who was himself an amateur astronomer. When Upjohn retired in 1944, he had been observing for thirteen years using binoculars, a 70-mm Goerz refractor, and a 5-inch Clark refractor. In need of a service to clean and repair his telescope eyepieces in 1944, he contacted the Perkin-Elmer company for help, and was referred (probably by AAVSOer Charles Elmer himself) to AAVSO Recorder Leon Campbell. Upjohn applied for AAVSO membership in 1946 and began reporting his variable star observations. Responding to commentary about the expense of publishing AAVSO reports, Upjohn, in 1949, offered to pay the cost of completing one of the reports. This offer led to his funding Campbell's *Studies in Long Period Variables* through an anonymous donation. Upjohn came to the aid of the AAVSO again in later years, but his 1949 donation was the most substantial and the most significant. He was named a Patron of the association in 1950. (Bennett 2014)

John J. Ruiz
(1894–1978)

by Tim Crawford

Ruiz, an electrical engineer, joined the AAVSO in 1947. He was a pioneer in the development of photoelectric photometry for amateur observers, and willingly shared his designs. He described his first photometer in "The Gremlins and My Photometer" (*Sky & Telescope*, December 1951). Ruiz was instrumental in publishing the *AAVSO Photoelectric Photometry Handbook* in 1956. Active with the PEP committee, he traveled nationally and internationally to promote photoelectric photometry. Ruiz was an inspiration to newcomers. He authored several refereed papers as well as having articles published in both *Sky & Telescope* and *Scientific American*. (Boss 1980)

Wayne M. Lowder
(1932–2003)

by Tim Crawford

Wayne M. Lowder, a radiation physicist, joined the AAVSO in 1949 while still a teenager. He earned a B.S. in physics from Harvard and did graduate work at Columbia. As one of the most active visual observers in AAVSO history, Lowder completed 208,630 observations, mainly with binoculars and his unaided eyes. His eyes were reportedly like a photometer; in later years he researched the relationship between visual and CCD-V magnitudes of the comparison stars used by the AAVSO. Not only was Lowder a leading observer but he was a leader in efforts during the 1960s and 1970s to bring the AAVSO into the "modern age." While Lowder studied variable star research efforts taking place around the world, he was impressed with the advances being made by the Russians in this field. Therefore, he spent much time searching Russian literature for work that would benefit the AAVSO membership. Lowder served as president (1993–1995) and then as treasurer (1999–2001). He authored a number of papers which appeared in *JAAVSO* between 1972 and 1999 and presented numerous papers at AAVSO Annual Meetings. (Mattei 2003a)

Frank J. De Kinder
(1892–1970)

by Tim Crawford

De Kinder, a construction cost estimator, joined the AAVSO in 1950. He was one of the early AAVSO solar observers and a participant in the nova search program. De Kinder served as president for the 1967–1969 term. As a resident of Canada and a member of the Royal Astronomical Society of Canada (RASC), he received that society's prestigious Chant Medal in 1955 for his astronomical work. De Kinder was one of the recognized developers of the RASC Montreal Centre's Observatory Observing program, having been a Centre member since 1934. (Anon. 1970b; Broughton 1994)

Curtis E. Anderson
(1927–1976)

by Michael Saladyga

Curtis Anderson joined the AAVSO in 1951 at age 24. He was a shipping clerk in Minneapolis, and had already spent five years as an amateur astronomer. He later resided in Coon Rapids, Minnesota. Through his first ten years with the AAVSO he made over 21,000 observations, but in 1961 he was stricken with multiple sclerosis. Anderson maintained his interest in variable stars nevertheless, and his yearly rate of observations increased over the next fifteen years, reaching a lifetime total of over 57,000 observations. During the time of his illness he helped Carolyn Hurless compile "inner sanctum" totals for her *Variable Views* newsletter, and he compiled the lifetime totals of AAVSO observers for the association's first fifty years. For his observing, historical compilations, and encouragement of young astronomers, Anderson received the 18th AAVSO Merit Award in 1965. As Hurless wrote, Anderson was "an inspiration as a faithful observer" to all who knew him, just as he was inspired by the good wishes of his fellow AAVSOers. (Hurless 1977)

Michel Daniel Overbeek
(1920–2001)

by Brian Fraser

Danie Overbeek's main variable star interest was the monitoring of cataclysmic variables for outbursts. He also observed many Mira variables and R CrB stars. He joined the AAVSO in 1953. Observing from Edenvale, near Johannesburg, South Africa, with an excellent home-made 12-inch Cassegrain, he contributed 292,711 observations to the AAVSO International Database. Less well-known is Danie's international reputation in grazing occultation observation. He was also a longtime, expert solar observer. His dedication and professional approach to observing earned him several prestigious awards, including the David Gill Medal of the Astronomical Society of Southern Africa (ASSA), and the 26th AAVSO Merit Award, and enormous respect from many quarters. He collaborated in many satellite-based astronomy projects with professionals. (Mattei and Fraser 2003)

Charles M. Good
(1904–1980)

by Tim Crawford

Good joined the AAVSO in 1956, serving as a council member from 1957 through 1968. He also served as president for the 1971–1973 term. His special interest was timing occulations and he was a longtime chairman of the AAVSO's lunar occultation committee. As a resident of Canada and a member of the Royal Astronomical Society of Canada (RASC), he received a Service Award medal in 1960 from the RASC for his many contributions. In 1981 the RASC Montreal Centre created the Charles M. Good Award to honor his memory as a principal contributor to the development of that Centre. (Williamson 1980; Broughton 1994)

Donald W. Engelkemeir
(1919–1969)

by Thomas R. Wiliams

Nuclear chemist Donald Engelkemeir joined the AAVSO in 1957. With a photoelectric photometer he designed and constructed, he observed an unusual stellar event, published in his paper "Photoelectric Observation of a Flare on AD Leonis" (Engelkemeir 1959). His observation of an Algol minimum on October 28, 1962, served as the basis for *Sky & Telescope* predictions for a number of years thereafter. Lewis Boss and John Ruiz invited Engelkemeir to join the AAVSO photoelectric photometry committee. Together, the committee published the first AAVSO PEP photometry handbook in 1962. Engelkemeir's photometer design served as the basis for the suggested instrument published in the handbook. (Anon. 1969)

Casper H. Hossfield
(1918–2002)

by Thomas R. Williams

Even up to the time of his death at age 84, New Jersey resident "Cap" Hossfield remained inventive and scientifically curious, characteristics that he exhibited throughout his 44-year membership in the AAVSO. Although he attained only a high school education, these characteristics informed his work career as a machinist and his other avocational career as a ham radio operator. For the AAVSO, Cap's interests were almost exclusively in the solar division, as chairman of the division from 1963 to 1980, and as editor of the *Sudden Ionospheric Disturbance Bulletin*. He designed and continued to improve the receiver and antenna for the SID receiver. Cap served fifteen years on council including one term as AAVSO President (1969–1970). Cap never failed to brighten the meetings he attended with his happy demeanor. (Feehrer 2003)

Carolyn J. Hurless
(1934–1987)

by Glenn Chaple

Carolyn Hurless was a music teacher by day and a dedicated variable star observer by night. A lifelong resident of Lima, Ohio, she joined the AAVSO in 1959. Under the tutelage of the legendary Leslie Peltier, Hurless became an accomplished and prolific variable star observer. In addition to several terms as council member, she served as second vice-president from 1967 to 1973. For twenty-two years she published the newsletter *Variable Views*. At the time of her untimely death in 1987, Carolyn Hurless had submitted 78,876 variable star observations, the highest total for a woman in the AAVSO. The AAVSO's Carolyn Hurless Online Institue for Continuing Education (CHOICE) is named in her honor. (Mattei 1987)

Carl A. Anderson
(1916–1994)

by Michael Saladyga

A native of Kansas City, Missouri, Carl Anderson settled in Manchester, New Hampshire, and was active in amateur radio and astronomy before he joined the AAVSO in 1960. Anderson was a member of the Royal Astronomical Society of Canada, and president of the Manchester (NH) Astronomical Society. He was a test engineer for a precision instrument manufacturer in Toronto and Manchester, and later was its president and treasurer. He was also a board member, trustee, and advisor for regional businesses and charities. Besides variable star observing, he participated in nova search, photographic photometry, occultation timing, and comet hunting, all of which he would introduce to members of his astronomical society. He eventually donated his 10-inch Cave reflector to the observatory at St. Anselm College in Manchester. He received the 19th AAVSO Merit Award in 1965 and served on the AAVSO council (1964–1967; 1974–1983; president 1979–1981). He was perhaps the first council member to express a need for bringing the Director's salary to a level comparable with other technical positions in the region. Also, at the urging of council member Tom Williams, Anderson called for a special session of the council in 1980 to plan an AAVSO Futures Study. (Harris 1994)

Arthur J. Stokes
(1918–2001)

by Tim Crawford

Art Stokes joined the AAVSO in 1962. He served several terms on the council, becoming president for the 1981–1982 term. An accomplished photoelectric photometry observer, Stokes served as chairman of the photoelectric photometry committee for over ten years. In later years he became an active radio monitor of solar sudden ionospheric disturbances and edited the AAVSO's *Solar Technical Bulletin*. He also invented a very low frequency receiver system for monitoring solar flares. Arthur published numerous refereed papers and was honored as the recipient of the 29th AAVSO Merit Award in 1987. (Mattei 2002)

Howard Joseph Landis
(1921–2014)

by Elizabeth O. Waagen

Howard Landis was born in Columbus, North Carolina, the youngest of five children. He served in the U.S. Army during World War II, and was an electronics technician for an airline for over thirty years. He joined the AAVSO in 1968, served on council (1975–1979), and was chair of the photoelectric photometry committee (1975–2003). He began publication of the *AAVSO Photoelectric Photometry Bulletin*, developed PEP reduction protocols, and maintained the PEP data archive. He contributed 1,675 PEP observations to the AAVSO International Database, and was advisor, helper, and mentor to amateur and professional astronomers. The AAVSO awarded an Honorary Membership to Howard Landis in October 2014. (Anon. 2015)

Theodore H. N. Wales
(1931–2003)

by Tim Crawford

Ted Wales, a 1952 Harvard graduate and investment banker, joined the AAVSO in 1975, his application endorsed by Dorrit Hoffleit of the Maria Mitchell Observatory. Wales was elected to the council in 1977 and became treasurer in 1979, serving in that role through 1998. Credited for his wise handling of the organization's finances during a difficult period in the 1980s, Wales significantly contributed to the AAVSO's survival and growth. He was well-known as a volunteer, always willing to help out wherever needed, whether it was working with data files or helping staff stain library shelves. He was a generous benefactor to the AAVSO through his annual donations, special gifts, and matching grants over the years. In 1991 he received the 33rd AAVSO Merit Award "in recognition of his loyalty and devotion to the Association, his untiring support in financial management and advice in over twelve years of service as AAVSO treasurer, and his many other contributions to the Association as a member of the council and a volunteer at Headquarters." (Mattei 2003b)

Martha Locke Hazen
(1931–2006)

by David B. Williams

Harvard astronomer Martha Hazen joined the AAVSO in 1975, served as president in 1992 and secretary, 1993 to 2004, and as a wise friend and counselor for director Janet Mattei. She also contributed effectively to the AAVSO's Futures Study Group. Hazen hosted many AAVSO meetings at the observatory's Phillips Auditorium. As curator of Harvard College Observatory's vast photographic plate archive for thirty-five years, she welcomed many variable star investigators, both professional and amateur, to use this invaluable resource. Her research focus was on variable stars in the previously neglected southern globular clusters. She was also a leader in expanding opportunities for women in astronomy. Hazen received the 37th AAVSO Merit Award in 2005. Her marriage to AAVSO member William Liller in 1959 ended in divorce in 1982. She married Bruce McHenry in 1991. (Anon. 2007; Williams and Willson 2007)

Louis Cohen
(1937–2013)

by Elizabeth O. Waagen

Lou Cohen was a computer engineer who designed compilers and operating systems, and a consultant engineer for software development. He was a recognized expert in a planning process for product development called "Quality Function Deployment," and published a book and consulted businesses on that subject after his retirement. Cohen joined the AAVSO in 2000 and served as treasurer from 2000 to 2006 and was a valued advisor to director Janet Mattei, interim director Elizabeth Waagen, and director Arne Henden. When he stopped observing, he donated his 30-cm telescope, CCD, and other equipment to the AAVSO; these are now part of the AAVSOnet Cohen/Menke Observatory in New Hampshire. Cohen was also a music composer, and taught and mentored in math, astronomy and music. (Anon. 2013a)

3. Name index

Note: Page numbers of names mentioned in passing within a particular biography are given in italics.

Abbott, Herbert Whipple, *95*
Anderson, Carl A., 98
Anderson, Curtis E., 97
Bancroft Jr., Harry C., *88*
Bappu, M. K. Vainu, *89*
Barnum, P. T., *89*
Bateson, Doris, *95*
Bateson, Frank M., 95
Boss, Lewis J., 91, *97*
Bouton, Tilton C. H., 88
Brashear, John A., *87*
Brocchi, Dalmiro F., 90
Buckstaff, Ralph N., 90
Campbell, Leon, *90, 91, 92, 95, 96*
Carpenter, Claude B., 94, *96*
Carpenter, Martha E. Stahr, 94
Chandler Jr., Seth C., 86, *86*
Chandra, Radha G., 89
Cilley, Morgan, 91, *94*
Cohen, Louis, 99
Cragg, Thomas A., *94*, 96
De Kinder, Frank J., 96
de Kock, Reginald P., 93
Elmer, Charles W., *89*, 90, *96*
Eisinge, Eise, *93*
Engelkemeir, Donald W., 97
Farnsworth, Alice H., *87*, 90
Fernald, Cyrus F., 94
Fernald, Emily, *94*
Ford, Clinton B., *94, 96*
Furness, Caroline E., *87*, 88
Godfrey, Charles C., 89
Good, Charles M., 97
Gott, Preston F., *94*
Gould, Benjamin A., *86*
Gray, Edward, 88
Haas, Phoebe W., 92
Halbach, Edward A., 93
Hamilton, Alexander, *94*
Hamilton, Mary C. S., *94*
Hamilton, Richard W., 94
Hazen, Martha L., *87*, 99
Heines, Neal J., 93
Henden, Arne A., *99*
Hiett, F. Lancaster, 94
Hindmann, J. V., *95*
Hoffleit, E. Dorrit, 92, *92, 98*
Hogg, Frank, *91*
Hogg-Priestley, Helen Sawyer, *87*, 91
Homer, Winslow, *89*
Hossfield, Casper H., 97
Houston, Walter S., *90*, 93
Hurless, Carolyn J., *97*, 98
Jenkins, Louise, *87*
Jones, Albert F. A. L., 95
Jones, Eugene H., 91
Jones, Fred, *91*
Kearons, William M., *92*
Kearons, Winifred C., 92
Kerr, J., *95*
Lacchini, Giovanni B., 88
Landis, Howard J., 98
Liller, William, *99*
Lowder, Wayne M., 96
Luizet, Michel, *86*
Luyten, Willem J., 89, *92*

Lyon, Margaret S., *89*
Mattei, Janet Akyüz, *95, 99*
Mayall, Margaret W., *92, 95*
McAteer, Charles Y., 87
McHenry, Bruce, *99*
Meek, Joseph, *93*
Mitchell, Maria, *87*
Olcott, Clara Hyde, 87
Olcott, William T., *87, 88, 89*
Overbeek, M. Daniel., 97
Parkhurst, John A., *91*
Payne-Gaposchkin, Cecelia, *91*
Peltier, Leslie P., 89, *98*
Perkin, Richard, *90*
Pickering, David B., 88
Pickering, Edward C., *88*
Plassmann, Joseph, *86*
Priestley, Francis, *91*
Reeves, Gertrude E. M., *95*
Reeves, Walter P., 95
Richter, Charles F., 89
Rosebrugh, David W., 93
Royer, Ron, *94, 96*
Ruiz, John J., 96, *97*
Sawyer, Edwin F., 86
Scanlon, Leo J., 92
Shapley, Harlow, *89, 91, 93*
Smith, J. Russell, 94, *94*
Stephansky Abbott, Helen M., 95
Stokes, Arthur J., 98
Swartz, Helen M., 88, *94*
Thomas, Helen M. L., 92
Upjohn, Lawrence N., 96
Upjohn, Uriah, *96*
Waagen, Elizabeth O., *99*
Wales, Theodore H. N., 98
Webb, Harold B., 91
Whitney, Mary W., 87, *88*
Williams, Thomas R., *98*
Witherell, Percy W., 90
Yalden, J. Ernest. G., 89, *90, 93*
Yendell, Paul S., 86, *86*
Young, Anne S., 87, *90, 91*
Young, Charles A., *87*
Young, Elizabeth, *87*

4. Sources

Biographical information for most subjects can be found in the published resources cited in the text and given in the reference list; additional information is from various sources discovered by the contributors, from the AAVSO archives, and in many cases from Williams and Saladyga (2011).

5. Acknowledgements

The editors of this paper thank the following contributors for their research and writing: Glenn Chaple of Townsend, Massachusetts; Tim Crawford of Arch Cape, Oregon; Brian Fraser of Henley-on-Klip, South Africa; Kate Hutton of Pasadena, California; Kristine Larsen of New Britain, Connecticut; Dee Sharples of Honeoye Falls, New York; Robert Stencel of Denver, Colorado; Matthew Templeton and Elizabeth O. Waagen of AAVSO Headquarters; and David B. Williams of Whitestown, Indiana.

References

Anon. 1924, *Pittsburgh Post*, October 20 (McAteer).
Anon. 1951, *Norwich (Connecticut) Bulletin*, May 12 (Hyde Olcott).
Anon. 1955, *Sky & Telescope*, **14** (October), 499 (Cilley).
Anon. 1959, *Norwalk (Connecticut) Hour*, May 25, 1 (Swartz).
Anon. 1960a, *Boston Globe*, October 2 (Farnsworth).
Anon. 1960b, *Sky & Telescope*, **20** (November), 262 (Farnsworth).
Anon. 1967, *Sky & Telescope*, **33** (March), 149 (Lacchini).
Anon. 1969, *Sky & Telescope*, **38** (July), 29 (Engelkemeir).
Anon. 1970a, *Boston Globe*, May 23 (Witherell).
Anon. 1970b, *Skyward: Mon. Newsl. Montreal Centre, Roy. Astron. Soc. Canada*, May, 3 (De Kinder).
Anon. 1980, *Sky & Telescope*, **59** (March), 209. (Fernald).
Anon. 1988, *AAVSO Newsletter*, No. 4, 11 (Rosebrugh).
Anon. 1993, *AAVSO Newsletter*, No. 10, 7 (C. Carpenter).
Anon. 1994a, *AAVSO Newsletter*, No. 12, 12 (Houston).
Anon. 1994b, *Sky & Telescope*, **87** (March), 11 (Houston).
Anon. 2007, *AAVSO Newsletter*, No. 35, 6 (Bateson; Hazen; Hoffleit).
Anon. 2013a, *AAVSO Newsletter*, No. 58, 9 (Cohen).
Anon. 2013b, *AAVSO Newsletter*, No 58, 9 (A. Jones).
Anon. 2015, *AAVSO Newsletter*, No. 64, 11 (Landis).
Bandyopadhyay, A., and Chakrabarti, R. 1991, *Indian J. Hist. Sci.*, **26**, 1 (Chandra).
Bennett, J. 2014, "A Brief History of the Upjohn Company" (http://www.pearce-bennett.freeserve.co.uk/ancestors/upjohn/weupjohn.htm) accessed April 2015 (Upjohn).
Bondy, H. L. 1955, *AAVSO Solar Bull.*, No. 108-109 (August-September), 2 (Heines).
Boss, L. J. 1980, *J. Amer. Assoc. Var. Star Obs.*, **9**, 48 (Ruiz).
Bracher, K. 2012, *J. Amer. Var. Star Obs.*, **40**, 24 (Young).
Broughton, R. P. 1994, *Looking Up. A History of the Royal Astronomical Society of Canada*, Dundurn Press, Toronto (De Kinder, Good, Hogg, Rosebrugh).
Cahill, M. J. 2012, *J. Amer. Assoc. Var. Star Obs.*, **40**, 31 (Hogg).
Callum, W. 1937, *Amateur Astron.*, **3** (October), 2 (Scanlon).
Campbell, L. 1946, *Pop. Astron.*, **54** (November), 443 (Pickering).
Christie, G. 2007, *J. Br. Astron. Assoc.*, **117**, 275 (Bateson).
Clement, C. 1993, *J. Amer. Assoc. Var. Star Obs.*, **22**, 83 (Hogg).
Ekvall, A. 1979-1980, *Community* (Univ. Wisconsin, Oshkosh), **2** (December-January), 10 (Buckstaff).
Engelkemeir, D. W. 1959, *Publ. Astron. Soc. Pacific*, **71**, 522.
Farwell, H. W. 1937, *Science*, **85** (April 2), 325 (Yalden).
Feehrer, C. E. 2003, *J. Amer. Assoc. Var. Star Obs.*, **31**, 171 (Hossfield).

Furness, C. 1922-1923, *Pop. Astron.*, **30** (December), 597, **31** (January), 25 (Whitney).

Glass, I. S. 1986, *Mon. Notes Astron. Soc. Southern Africa*, **45**, 113 (de Kock).

Harris, J. 1994, *The Observer: Newsl. New Hampshire Astron. Soc.*, February, 1 (C. A. Anderson).

Hoffleit, D. 2000, *J. Amer. Assoc. Var. Star Obs.*, **28**, 40 (Thomas).

Hoffleit, D. 2002, *Misfortunes as Blessings in Disguise: The Story of My Life*, AAVSO, Cambridge, MA (Hoffleit).

Hurless, C. 1977, *J. Amer. Assoc. Var. Star Obs.*, **6**, 9 (C. E. Anderson).

Hurless, C. 1980a, *J. Amer. Assoc. Var. Star Obs.*, **9**, 32 (Peltier).

Hurless, C. 1980b, *Sky & Telescope*, **60** (August), 104 (Peltier).

Ingham, J. A. 1937, *Variable Comments*, **3**, 52 (Yalden).

Larsen, K. 2012a, *J. Amer. Assoc. Var. Star Obs.*, **40**, 44 (Hoffleit).

Larsen, K. 2012b, *J. Amer. Assoc. Var. Star Obs.*, **40**, 51 (Stahr Carpenter).

Luyten, J. R. 1995, *Bull. Amer. Astron. Soc.*, **27**, 1480 (Luyten).

Makemson, M. W. 1936, *Pop. Astron.*, **44** (May), 233 (Furness).

Mattei, J. A. 1987, *J. Amer. Assoc. Var. Star Obs.*, **16**, 35 (Hurless).

Mattei, J. A. 2002, *AAVSO Newsletter*, No. 27, 3 (Stokes).

Mattei, J. A. 2003a, *AAVSO Newsletter*, No. 29, 9 (Lowder).

Mattei, J. A. 2003b, *AAVSO Newsletter*, No. 29, 9 (Wales).

Mattei, J. A., and Fraser, B. 2003, *J. Amer. Assoc. Var. Star Obs.*, **31**, 65 (Overbeek).

Mayall, M. W. 1955, *Sky & Telescope*, **14** (March), 189 (Elmer).

Mundt, C. S. 1920, death notice, AAVSO, Cambridge, MA (Gray).

Peltier, L. C. 1979, *J. Amer. Assoc. Var. Star Obs.*, **8**, 71 (Fernald).

Pickering, D. B. 1937, *Pop. Astron.*, **45** (June-July), 296 (Yalden).

Rosebrugh, D. W. 1937, *Amateur Astron.*, **3**, 1 (E. Jones).

Samolyk, G. 2011, *J. Amer. Assoc. Var. Star Obs.*, **39**, 141 (Halbach).

Searle, A. 1914, *Pop. Astron.*, **22** (May), 271 (Chandler).

Smith, J. R. 1938, *Amateur Astron.*, **4** (May), 1 (Smith).

Smith, J. R. 1959, *Sky & Telescope*, **18** (June), 431 (Smith).

Smith, J. R. 1973, *Sky & Telescope*, **45** (April), 250 (Smith).

Taibi, R. J. 2004, *WGN, J. Int. Meteor Organization*, **32**, 87 (Sawyer).

Toone, J. 2011, *Br. Astron. Assoc. Var. Star Sect. Circ.*, No. 149, 8 (Cragg).

Waldo, G. 1928, *Pop. Astron.*, **36** (April), 228 (Godfrey).

Wilford, J. N. 1985, *New York Times*, October 1.

Williams, T. R. 1991, *J. Amer. Assoc. Var. Star Obs.*, **20**, 18 (Haas).

Williams, T. R. 2007, in *Biographical Encyclopedia of Astronomers*, T. T. Hockey, V. Trimble, T. R. Williams (eds.), Springer, New York, 461 (Halbach).

Williams, T. R., and Saladyga, M. 2011, *Advancing Variable Star Astronomy: The Centennial History of the American Association of Variable Star Observers*, Cambridge University Press, Cambridge.

Williams, T. R., and Willson, L. A. 2007, *Bull. Amer. Astron. Soc.*, **39**, 1064 (Hazen).

Williamson, I. K. 1980, *Skyward: Mon. Newsl. Montreal Centre, Roy. Astron. Soc. Canada*, November-December, 8 (Good).

Some Personal Thoughts on TV Corvi

David H. Levy
Jarnac Observatory, P.O. Box 895, Vail, AZ 85641; doveed@sharingthesky.org

Received April 7, 2015; accepted June 11, 2015

Abstract As part of the AAVSO's role in celebrating the United Nations' International Year of Light, I have been asked to prepare a brief retrospective on my interest in Clyde Tombaugh's star, TV Corvi. Because of the clever light pollution ordinances that have governed the night sky surrounding the area around Tucson, Arizona, and the International Dark Sky Association, our Jarnac Observatory has been blessed with a dark night sky that often permits observations down to 19th magnitude, the star's suspected minimum magnitude. Preparing this article has also helped me to understand that variable star observing is not just science; it is community. My own understanding of the behavior of Tombaugh's Star is gathered from my long friendship with Clyde Tombaugh, discoverer of Pluto and the scientist who opened the door to the Kuiper Belt to other AAVSO observers over many years, Steve Howell, from the Planetary Science Institute, who alerted me to the possibility that one component of TV Crv is a brown dwarf, and the pure joy of being able to observe this faint variable star under a dark sky.

1. Introduction

Recently Dr. Stella Kafka, newly appointed Director of the AAVSO, asked me to write an article for *JAAVSO* about TV Corvi, my favorite variable star. One thing I learned long ago is that when a Director, particularly of the AAVSO, asks me to write something, the best thing to do is to drop whatever it is that I am doing and fulfill her request. My own concern for this particular star dates back almost thirty years to February 9, 1986, the perihelion date of Halley's comet. Sitting in the basement of Lowell Observatory, I was studying the original photographic logs of Clyde Tombaugh in preparation for a biography I was writing about him. On the plate exposed 10 January 1931 was circled the trailed image of a comet. I spent years trying to substantiate Clyde's images, but found nothing. Even though Clyde's telescopes recorded several images of this object, the International Astronomical Union's Central Bureau for Astronomical Telegrams would not announce it unless images from other observers could be found. The comet was rediscovered in 2012 on images taken by the Tenagra Observatories, and is now known as Comet 274P/Tombaugh-Tenagra.

The search for comets did not end there, however. During the summer of 1987 I retuned to Lowell and checked every one of Clyde's planet search plates. (Although the search for trans-Neptunian planets was what the search began as, after Pluto was discovered the search was expounded to a "trans-Saturnian planet" program, and this is how Clyde always referred to it.) This time I uncovered evidence of Clyde's discovery of a single star annotated:

> Nova. 1 nova suspect "T 12" near southwest corner of plate, magnitude about 12, confirmed well on Cogshall [telescope] plates of MARCH 22. No trace of object on 13-inch plates of March 20 and 17, 1931. The image is exactly deformed, like the other star images in the neighborhood. Evidently a very remarkable star to rise from 17 or fainter to 12 in 2 days time.... This object was discovered on May 25, 1932, at 11:00 AM. (Tombaugh 1931; Levy 1991)

Sixty years after the fact, Clyde was still alive and remembered well the moment of discovery. "It was a definitely a real star," the discoverer of Pluto told me over the phone, adding that its image was slightly deformed just like the surrounding stars near the edge of the plate he had taken. He knew it was a "temporary star" as he called it, because it did not appear on either of the other plates he had exposed of the same region. Brian Skiff, a staff astronomer there, suggested that I search some plate archives for other images of this star that could confirm its existence.

2. Confirming Tombaugh's star

On September 11, 1989, therefore, I visited the famous plate stacks at Harvard College Observatory, just a long block down the road from the AAVSO's old headquarters on Concord Avenue. Over three days, I searched through 260 patrol plates, probably the entire collection that HCO had containing the position of the star Clyde had discovered 58 years earlier. The search period spanned a long period of time, from 1930 to 1988. The search yielded nine additional outbursts of what I concluded had to be an SS Cygni-type dwarf nova.

Armed with this evidence, I walked across the lawn to Brian Marsden's office in an adjoining building. He looked at the list of outbursts I presented to him, then back at me. "I agree this is interesting," he said, "but I am not going to announce it yet."

"Why not?" I asked.

"Because," Brian answered sagely but with a grin and a wink, "you are an amateur astronomer." I took a couple of deep breaths, then prepared to say something less than friendly. Brian then added, "If you were a professional astronomer, you'd have to apply for telescope time, and probably you wouldn't bother with it. But as an amateur with a beautiful 40-centimeter reflector capable of discovering comet after comet, you can keep a visual watch on the star's position every night. When you next see the star in outburst, which I don't doubt would someday occur, then I will announce it as a current item."

Thus, in November 1989 when Corvus began to make its appearance in the predawn sky after solar conjunction, I began daily observations of the field.

On March 22, 1990, after giving a lecture in Florida, I checked the region using one of Don Parker's telescopes. The following night, back home, I used my own 40-cm reflector and saw a new star of magnitude 13.6 where nothing had been before!

Since that memorable night I have seen several further outbursts of this star, which Steve Howell of the Planetary Science Institute determined to be not only a high galactic latitude cataclysmic variable star (most such stars lie near the galactic plane), but also that it consists of a white dwarf and a brown dwarf that orbit each other in an area smaller than the Sun in a period under two hours (Planetary Science Institute 2005; Wood et al. 2011). One outburst in 2005 deserves particular note. The star was apparently just beginning its rise to maximum when I recorded it. I decided to take repeat exposures every thirty minutes for the rest of the night. My sequence recorded a series of images showing this beautiful star on its way to maximum, and at the AAVSO spring meeting a few months later I played the animation, actually "TV Corvi: The motion picture," during the paper session. After a few showings I went to shut it off, but the audience refused to allow this. Thus I had to continue the animation for the remainder of my paper. Incidentally, this episode is one of the reasons I love going to AAVSO meetings. (Another episode, that had nothing to do with TV Crv, took place during an evening observing session during a spring meeting at our Jarnac Observatory. I casually asked if anyone would be interested in seeing my small collection of old blueprint charts; within a minute the whole crowd was gathered round, admiring the way we used to do variable stars.)

3. Tombaugh's star and the community of variable star observers

Of all the outbursts I have seen since 1990, four have occurred near the date of my first one, March 23, which was coincidentally the date of Clyde's first detection back in 1931. The one on March 23, 2000, was so special to me that I informed then-director Janet Mattei about it by telephone. It meant so much to her that I called because it brought to the forefront her wish to see the human side of variable stars. And on that particular evening, she did. I have presented a paper and co-written other short pieces about this star, and have discussed it on countless occasions with many astronomers both amateur and professional (Levy et al. 1990; Levy 2000).

More important, Tombaugh's star has a unique role to play in the history of the AAVSO and in its many decades of outreach. It is important because it reminds us of one of the most important astronomical observers in the twentieth century. It is important because it suggests the existence of a stellar type that is very, very small; much smaller than our Sun and whose components might be not much larger than Jupiter. And for me, it reminds me of some interesting times that have happened in my life.

4. Superoutbursts for a super star

We now understand that TV Crv is an SU-Ursae Majoris variable: a cataclysmic variable whose outbursts come in two varieties, normal outbursts and superoutbursts, and that exhibits superhumps (small periodic variations related to the length of the orbital period) during superoutbursts. Outbursts occur when gas that is gathering in the accretion disk reaches a certain temperature, the viscosity in the disk changes, and the gas collapses onto the brown dwarf. As gravitational potential energy is released, the system brightens exponentially. This particular star's outbursts apparently result when the accretion disk surrounding the smaller star becomes unstable.

TV Crv (Figure 1) is a special type of SU UMa variable in that its superoutbursts come in two varieties—one with an uninterrupted rise to maximum, and one with a partial rise, slight decline, then full rise to maximum. This latter type is the result of a precursor normal outburst which happens to affect the disk in a way that triggers the superoutburst (Uemura et al. 2005).

For examples, we can revisit the superoutburst of 2001 18 February, during which the first recorded visual magnitude observation was 12.9 (Figure 2). In this event, TV Crv went into its superoutburst phase without warning; the preceding night the star was typically fainter than magnitude 14.6—there was no precursor in this superouburst.

The 2004 June 4 superoutburst was associated with a precursor. It appeared to begin as a normal outburst. TV Crv brightened from its quiescent state to about magnitude 13.7

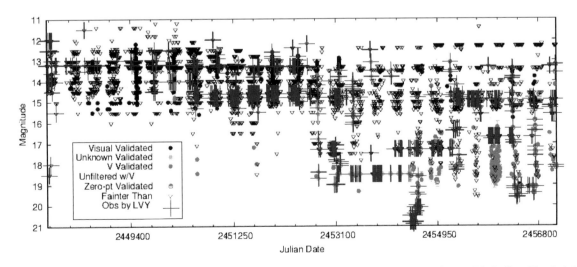

Figure 1. TV Crv, 17 November 1989–4 June 2015 (JD 2447848–2457178). Data from AAVSO International Database (AID). Earliest, historical data in AID (February 1930–April 1981, not shown) were digitized by D. Levy from the Harvard College Observatory plate collection and the Palomar Observatory Sky Survey.

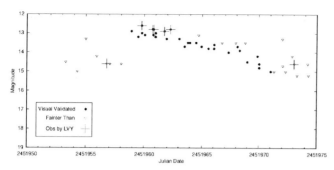

Figure 2. TV Crv, 9 February–6 March 2001 (JD 2451950–2451975). Data from AAVSO International Database.

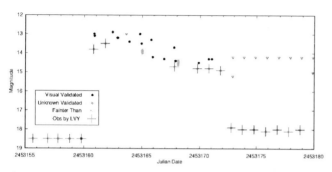

Figure 3 TV Crv, 29 May–23 June 2004 (JD 2453155–2453180). Data from AAVSO International Database.

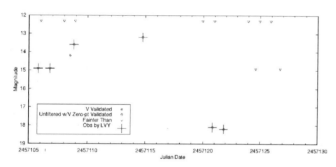

Figure 4. TV Crv, 23 March–17 April 2015 (JD 2457105–2457130). Data from AAVSO International Database.

over a few hours, and then, after a slight fading, continued brightening in a superoutburst, reaching maximum 1.7 days later (Uemura *et al.* 2005) (Figure 3).

In the superoutburst of 27 March 2015 (Figure 4), which took place as I was writing this paper, TV Crv was still at maximum when I observed it again on 2 April. It subsequently returned to minimum by May 17.

5. TV Crv: its astronomical and personal significance

Why exactly is TV Corvi, or Tombaugh's star, my favorite variable? This is not a hard question to answer. Every time I observe either the star itself or its field, I am reminded of my close friendship with Clyde Tombaugh. Most people know Clyde only for his discovery of Pluto, connected today with the continuing arguments over its status. Years ago Steve Howell told me that he considered TV Corvi to be Tombaugh's most significant discovery, far more so than his primary discoveries of the Kuiper Belt. As I was now devoting most of my observing hours to comets, it seemed appropriate to observe its field every night during its season to determine its outburst frequency. Although, as a cataclysmic variable, its outbursts cannot be predicted, the outbursts have the unlikely habit of occurring roughly once each year. March appears to be the favored month, and on four occasions the outbursts have taken place either on March 23, or have been in progress on that date or slightly after. Besides being the date on which more than two outbursts have been detected, March 23 is also the date marking the discovery of my most important comet, Shoemaker-Levy 9 in 1993, and it is the day I married Wendee in 1997. More recently, one of the telescopes I use for my nightly comet search is named "Clyde" not for his discoveries, but for the personality of the man: his love of science, his sense of humor, and his ubiquitous and unforgettable puns. All these things are rooted in this unusual cataclysmic variable, TV Crv. This wonderful pairing of two tiny stars has made a personal and continuing involvement in my life that I will not soon forget.

References

Levy, D. H. 1990, *Publ. Astron. Soc. Pacific*, **102**, 1321.
Levy, D. H. 1991, *Clyde Tombaugh: Discoverer of Planet Pluto*, Univ. Arizona Press, Tucson, 189.
Levy, D. H. 2000, *J. Amer. Assoc. Var. Star Obs.*, **28**, 38.
Planetary Science Institute. 2005, "TV Corvi" (http://www.psi.edu/about/staff/lbleamas/otw/star.html).
Tombaugh, C. 1931, plate envelope notes for March 22, 1931, Lowell Observatory, Flagstaff, AZ.
Uemura, M. *et al.*, 2005, *Astron. Astrophys.*, **432**, 261.
Wood, M., Still, M. D., Howell, S. B., Cannizzo, J. K., and Smale, A. P. 2011, *Astrophys. J.*, **741**, 105.

Abstracts of Papers and Posters Presented at the 103rd Annual Meeting of the AAVSO, Held in Woburn, Massachusetts, November 6–8, 2014

General Paper Session Part I

Betelgeuse Period Analysis Using VSTAR

Frank Dempsey
R. R. #1, 3285 Sideline 20, Locust Hill, ON L0H 1J0, Canada; cosmicfrank99@gmail.com

Abstract Betelgeuse was studied using the VSTAR software package and analysis of the observations in the AAVSO International Database. Period analysis derived a period of 376 days, in comparison with literature periods of 420 days using satellite UV data but significantly different from the VSX period of 2,335 days. The unique set of PEP observations of this star is also shown, and the advantage of PEP Johnson V observations is shown in comparison with the visual observations.

EE Cep Winks in Full Color

Gary Walker
21 Ashley Lane, P. O. Box 26, North Groton, NH 03266; bailyhill14@gmail.com

Abstract We observe the long period (5.6 years) Eclipsing Binary Variable Star EE Cep during its 2014 eclipse. It was observed on every clear night from the Maria Mitchell Observatory as well as remote sites for a total of 25 nights. Each night consisted of a detailed time series in BVRI looking for short term variations for a total of >10,000 observations. The data was transformed to the Standard System. In addition, a time series was captured during the night of the eclipse. These data provide an alternate method to determine Time of Minimum to that traditionally performed. The TOM varied with color. Several strong correlations are seen between colors substantiating the detection of variations on a time scale of hours. The long term light curve shows five interesting and different phases with different characteristics.

Transient Pulsation of Sirius

Kangujam Yugindro Singh
Irom Ablu Meitei
Address correspondence to K. Y. Singh, Department of Physics, Manipur University, Canchipur, Imphal West District, 795003, India; yugindro361@gmail.com

Abstract A photometric study of Sirius during the last week of January 2014 has revealed that the star shows transient pulsation. Observations of Sirius were taken with an integration period of ten seconds in B and V bands using a SSP3 photometer attached to a Celestron CGE1400 telescope. During the observations taken on the night of 30 January, 2014, the B-band magnitude remained almost unchanged; the difference in the maximum and minimum magnitude in B-band was 0.13. However, it was found that the V-band magnitude changed appreciably, decreasing by values up to more than 4. The pulsation in the V-band was so rapid that the V-band magnitude changed sometimes by more than 4 magnitudes in a short period of 10 seconds. Such a transient change in the properties of Sirius cannot be accounted for by the eclipsing binary phenomenon. Disruptive binary interactions such as mass transfer between Sirius A and B might account for such a powerful transient phenomenon.

η Carinae Continues to Evolve

John C. Martin
University of Illinois at Springfield Observatory, One University Plaza, MS HSB 314, Springfield, IL 62704; jmart5@uis.edu

Abstract η Carinae affords us a unique opportunity to study the pre-supernova evolution of the most massive stars. For at least the last half century, it has maintained a 5.5-year spectroscopic cycle that culminates with abrupt decreases in the strong stellar wind emission features. Over the last 15 years, the star has brightened at an accelerated rate and altered its spectrum, in addition to the spectroscopic cycle, indicating an ongoing change in state. We present Hubble Space Telescope spectroscopy and synthetic photometry from the most recent spectroscopic event (2014.5) that shows notable differences with past events and provides clues to the on-going evolution of the star.

The Trend in the Observation of Legacy Long Period Variable Stars (poster)

Robert Dudley
5613 River Road, Arlington, VT 19046; greyowlvermont@live.com

Abstract A decrease in the number of observers of the Legacy Long Period Variable Stars has been noted by the AAVSO. Amongst the observing community there is the perception that observers collecting digital data are making up for this gap. Data from the annual Director's report (2002–2013) and the AAVSO International Database (AID) for the years 1993, 2003, and 2013 were analyzed. For the period of 2002 to 2013 the total number of observers remained fairly constant (816±97) with a large bump in 2011. The number of visual observations has slowly declined since 2007 though there has recently been an increase in the number of observations. From the AID data the number of observations reached a maximum in 2003 and has slowly declined afterwards. These trends as well as other information gleaned from the data will be present and discussed.

Analysis of Hα Lines in ε Aurigae Post-eclipse (poster)

Shelby Jarrett
Cybil Foster
Address correspondence to S. Jarrett, 308 N Cedar Bluff Drive, Valmeyer, IL 62295; sjarr2@uis.edu

Abstract ε Aurigae is an eclipsing binary star system located in the constellation Auriga. The primary is an F type star that is eclipsed every 27 years by a large disk of material. Campaigns in the past have focused on the photometry and spectroscopy during the eclipse without much attention to the primary outside of the eclipse. Spectra taken throughout the year following the end of the last eclipse showed continued changes in the spectrum of the primary star outside of the eclipse. We seek to develop a model and comprehension of how the primary spectrum changes independent of the eclipse in order to establish a better understanding of the secondary and its influence on the spectrum during the eclipse. We will analyze medium-resolution spectra covering most of the visual range from the end of the 2011 eclipse to the present for significant patterns, focusing primarily on spectral regions and features that have been previously discovered in the interpretation of the eclipses.

Discovery of Five Previously Misidentified BY Draconis Stars in ASAS Data (poster)

Jessica Johnson
Department of Physics and Earth Sciences, Central Connecticut State University, 1615 Stanley Street, New Britain, CT 06050; johnsonj@my.ccsu.edu

Kristine Larsen
Department of Physics and Earth Sciences, Central Connecticut State University, 1615 Stanley Street, New Britain, CT 06053; Larsen@ccsu.edu

Abstract This work is a continuation of an ongoing project first presented at the 2013 Annual Meeting of the AAVSO. The original poster introduced a spreadsheet of 3,548 computer-classified candidate Cepheid variable stars in the ASAS (All Sky Automated Survey) photometry data, a data set that was known to contain many false positive identifications. Previous work by Patrick Wils suggested that BY Draconis stars (spotted K and M dwarfs) were an important source of misidentifications in this sample. The authors have undertaken a project to systematically identify previously unknown BY Draconis stars in this data set. This is initially done through the investigation of the stars' known physical properties (for example, from infrared photometry [2MASS] and proper motion [PPMXL] data). An analysis of light curves and phase plots is the final step in identifying BY Draconis stars, for example searching for characteristic changes in mean magnitude and amplitude. Thus far five previously unknown BY Draconis stars have been identified through this process.

AAVSO and the International Year of Light (poster)

Kristine Larsen
Department of Physics and Earth Sciences, Central Connecticut State University, 1615 Stanley Street, New Britain, CT 06053; Larsen@ccsu.edu

Abstract The United Nations General Assembly has officially designated 2015 to be the International Year of Light (IYL). Modeled in part on the earlier International Year of Astronomy (IYA), this cross-disciplinary, international educational and outreach project will celebrate the importance of light in science, technology, cultural heritage, and the arts. It ties in with several important anniversaries, such as the 1000th anniversary of the publication of Ibn Al Haythem's "Book of Optics," the 150th anniversary of Maxwell's equations of electromagnetism, the centenary of Einstein's General Theory of Relativity, and the 50th anniversary of the discovery of the Cosmic Microwave Background Radiation. Because variable stars are defined as such due to the variability of the light we observe from them, all of the AAVSO programs, regardless of type of variable or instrumentation (eye, DSLR, PEP, or CCD), have natural tie-ins to the study of light. This poster will highlight a number of specific ways that AAVSO members and the organization as a whole can become intimately involved with this unique outreach opportunity.

Precision Photometry of Long Period Variable Stars: Flares and Bumps in the Night (poster)

Dale Mais
14690 Waterstradt Street, Marcellus, MI 49067; dale.mais@mpiresearch.com

Abstract Mira variable stars are a broad class of stars, which encompass spectroscopic classes of type M, S, and C. These stars are closely related in terms of their long term variability, position on the Hertzsprung-Russell diagram, their intermediate mass (from ~0.8 to ~8 solar mass), and the fact that class M evolves into the S and C type stars as certain stages of shell burning around the core proceeds. Recently, evidence has accumulated to suggest that Mira variables may go through flare-up stages which result in brightening on the order of several tenths of a magnitude or more and may last hours to days in length. Very little is known about these events, indeed it is not clear that these events are real. In order to address the reality of these events, we established an automated acquisition/analysis of a group of 108 Mira variables in order to obtain the densest coverage of the periods to better constrain the potential flare-ups. Telescope control scripts were put in place along with real time analysis. This allowed for unattended acquisition of data on every clear night, all night long, in the V, R, and I photometric bands. In addition, during the course of the night multiple determinations were often obtained for a given star. The light curves of many of the program stars show a Cepheid-like bump phenomenon, however these appear on the ascending part of the light curve. In general, these bumps appear in longer period Miras (>350 days). Bumps are not obvious or easily seen in *visual* data records, although slope changes during rising phase are seen in some cases. So far, greater than 100,000 magnitude determinations have been obtained, many closely spaced in time. This should help to further constrain the potential occurrences of flare-up events.

Transformation: Adjusting Your Data to the Standard Photometric Framework (poster)

George Silvis
194 Clipper Road, Bourne, MA 02532; gasilvis@gmail.com

Abstract This year the AAVSO made an effort to present tools to the membership and observership to help them get their CCD data transformed. My poster offers a description of what transforming does, including a visualization of how it adjusts your data. And it shows the new TransformApplier application in action.

The Eggen Card Project (poster)

George Silvis
194 Clipper Road, Bourne, MA 02532; gasilvis@gmail.com

Abstract At the 2013 meeting we kicked off the Eggen Card project. This project was to make the huge collection of photometric observations made by Olin Eggen accessible to researchers. My poster this year is to report progress and encourage more members and observers to participate as volunteers.

Visual Spectroscopy of R Scuti (poster)

Lucian Undreiu
Andrew Chapman
Address correspondence to L. Undreiu, 121 Beverly Avenue, Wise, VA 24293; lundreiu@hotmail.com

Abstract We are currently conducting a visual spectral analysis of the brightest known RV Tauri variable star, R Scuti. The goal of our undergraduate research project is to investigate this variable star's erratic nature by collecting spectra at different times in its cycle. Starting in late June of 2014 and proceeding into the following four months, we have monitored the alterations in the spectral characteristics that accompany the progression of R Sct's irregular cycle. During this time, we were given the opportunity to document the star's most recent descent from maximum brightness V~5 to a relatively deep minimum of V~7.5. Analysis of the data taken during the star's period of declining magnitude has provided us with several interesting findings that concur with the observations of more technically sophisticated studies. Following their collection, we compared our observations and findings with archived material in the hopes of facilitating a better understanding of the physical state of RV Tauri stars and the perplexing nature of their evolution. Although identification of the elements in the star's bright phase proved to be challenging, documenting clear absorption features in its fainter stage was far less difficult. As previously reported in similar studies, we identified prominent TiO molecular absorption bands near R Sct's faintest state, typical of mid-M spectral type stars. In addition to these TiO absorption lines, we report the presence of many more metallic lines in the spectral profiles obtained near star's minimum. Supportive of previously published hypotheses regarding the causation of its variability, we observed significant variation in the star's spectral characteristics throughout different phases of its cycle. We are hopeful that our observations will make a meaningful contribution to existing databases and help advance our collective understanding of RV Tauri stars and their evolutionary significance.

General Paper Session Part II

Parallel Group and Sunspot Counts from SDO/HMI and AAVSO Visual Observers

Rodney Howe
3343 Riva Ridge Drive, Fort Collins, CO 80526; ahowe@frii.com

Jan Alvestad
Vigdelsvn. 637B, 4054 Tjelta, Norway; jan@solen.info

Abstract Creating group and sunspot counts from the SDO/HMI detector on the Solar Dynamics Observatory (SDO) satellite requires software that calculates sunspots from a "white light" intensity-gram (CCD image) and group counts from a filtered CCD magneto-gram. Images from the satellite come from http://jsoc.stanford.edu/data/hmi/images/latest/ Together these two sets of images can be used to estimate the Wolf number as $W = (10g + s)$, which is used to calculate the American Relative index. AAVSO now has approximately two years of group and sunspot counts in its SunEntry database under SDOH observer Jan Alvestad. It is important that we compare these satellite CCD image data with our visual observer daily submissions to determine if the SDO/HMI data should be included in calculating the American Relative index. These satellite data are continuous observations with excellent seeing. This contrasts with "snapshot" earth-based observations with mixed seeing. The SDO/HIM group and sunspot counts could be considered unbiased, except that they show a not-normal statistical distribution when compared to the overall visual observations, which show a Poisson distribution. One challenge that should be addressed by AAVSO using these SDO/HMI data is the splitting of groups and deriving group properties from the magneto-grams. The filtered CCD detector that creates the magento-grams is not something our visual observers can relate to, unless they were to take CCD images in H-alpha and/or the Calcium spectrum line. So, questions remain as to how these satellite CCD image counts can be integrated into the overall American Relative index.

Going Over to the Dark Side

David Cowall
20361 Nanticoke Drive, Nanticoke, MD 21840; cowall@comcast.net

Abstract This is the tale of my continuing journey transforming from a visual to a CCD photometrist. It is my hope that sharing my experiences will help and encourage others to consider taking the same path. It has been hard, but fun—a wonderful opportunity as a newly retired physician to expand my horizons. However, my brain did have to make the switch from Biology to Physics. The major barrier that concerned me was cost, but change itself was also a challenge. Other issues included dealing with the complexity of technical systems and a myriad of details. My solution was to be patient and think small to insure success and then build upon all those little victories. The pedagogical component of this project was critical as well. It began with a good mentor and continued via networking with other members at meetings, taking CHOICE courses, and most importantly: practice, practice, practice. Each plateau suggested many new possibilities. I think "The Force" is now with me! The adventure continues.

Photometry Transforms Generation with PTGP

Gordon Myers
5 Inverness Way, Hillsborough, CA 94010; GordonMyers@hotmail.com

Ken Menzies
318A Potter Road, Framingham, MA 01701; kenmenstar@gmail.com

George Silvis
194 Clipper Road, Bourne, MA 02532; gasilvis@gmail.com

Barbara Harris
1600 South State Road 415, New Smyrna Beach, FL 32168; barbharris1@hughes.net

Abstract Historically the development of photometry transformation coefficients required extensive manual effort and the use of large spreadsheets. A new release—version 5.0—of the Photometry Transformations Generation Program (PTGP) achieves the goal of generating transformation coefficients without the use of spreadsheets—saving considerable time and ensuring data accuracy. PTGP version 5.0 works directly with the AAVSO Variable Star Plotter (VSP) to retrieve the most recent standard star reference magnitudes (currently for M67 and NGC 7790). It then processes instrument magnitude file(s) downloaded from VPHOT or exported from AIP4WIN or MAXIM. Either AUID or "Boulder" star identifications can be used for AIP4WIN and MAXIM. When using VPHOT data or "Boulder" star identifications, PTGP determines the AUID names for each of the reference standard stars. All standard transforms are calculated. Plots of each transform's data can be reviewed, and individual star observations added/deleted. Transform sets can be saved for further use. Transform sets can be compared and selected sets averaged. The averaged sets can be exported in a file format compatible with the AAVSO TA tool. The presentation will provide a brief overview and demonstration of the tool. It will also discuss the implications of using PYTHON for the development—both benefits and potential problems. The program runs on both PCs and Macs. A subsequent presentation will discuss the use of VPHOT and PTGP to generate transforms and the testing of the impacts of varying key VPHOT and PTGP parameters.

Using VPHOT and PTGP to Generate Transformation Coefficients

Ken Menzies
318A Potter Road, Framingham, MA 01701; kenmenstar@gmail.com

Gordon Myers
5 Inverness Way, Hillsborough, CA 94010; GordonMyers@hotmail.com

Abstract The AAVSO web site hosts two useful tools, VPHOT and PTGP, to help develop your transformation coefficients. They can be used together to simplify a tedious process involving standard comparison star selection, image reduction, and spreadsheet analysis. The process necessary to generate Transformation Coefficients involves: (1) measurement of instrumental magnitudes of Standard Comparison Stars, (2) measurement of instrumental magnitudes for a set of images and filters (UBVRI), and (3) downloading of comparison star magnitude files in a standard format. The subsequent process involves: (4) importing the set of standard comparison star magnitude files into PTGP, and (5) the automatic calculation of transformation coefficients and transformation plots. During testing in VPHOT and PTGP, simple steps and alternatives have been identified to generate accurate transform coefficients. The first VPHOT step is the obvious need to upload multiple images in multiple standard filters to VPHOT. The second step is to open each image, overlay all standard comparison stars (currently M67 or NGC 7790), view the photometry table, and download the comparison star magnitude data files. Two alternatives involve either the use of individual image files (for example, 4B, 4V, 4R; 4I) or the stacking of all images for each filter (for example, 1B, 1V, 1R; 1I). The latter improves the SNR. In PTGP after the selection of telescope, standard fields, and data reduction software, one selects all comp star magnitude files. The "Calculate Transform Set" button calculates the applicable filter magnitude and color index coefficients. Testing of this process included an evaluation of the impact of several alternatives including: (1) choice of individual filter images or stacked filter images, (2) choice of all standard comparison stars or a selected subset of standard comparison stars, and (3) choice of a minimum SNR. Each of these alternatives affects the transformation coefficients to a small extent.

General Paper Session Part III

Observational Activities at Manipur University, India

Kangujam Yugindro Singh
Irom Ablu Meitei
Salam Ajitkumar Singh
Rajkumar Basanta Singh
Address correspondence to K. Y. Singh, Department of Physics, Manipur University, Canchipur, Imphal West District, 795003, India; yugindro361@gmail.com

Abstract We have innovatively designed and constructed three observatories each costing a few hundred USD for housing three small Schmidt-Cassegrain type telescopes, namely, Celestron CGE925, Celestron CGE1400, and Meade 12-inch LX200GPS. These observatories are completely different in design and are found to be perfectly usable for doing serious work on astronomical observation and measurements. The observatory with the Celestron CGE1400 telescope has been inducted, since January 2012, as one of the observatories of the international "Orion Project" headquartered at Phoenix, Arizona, which is dedicated for photometric and spectroscopic observations of five bright variable stars of the Orion constellation namely, Betelgeuse (α Ori), Rigel (β Ori), Mintaka (δ Ori), Alnilam (ϵ Ori) and Alnitak (ζ Ori). Using this observatory, we have been producing BVRI photometric data for the five stars of the Orion project. The other observatory with the Meade 12-inch LX200GPS telescope is being inducted into service for CCD photometric study of SU UMa stars in connection with implementation of a project funded by the Indian Space Research Organization (ISRO). In the present paper, we would like to describe our self-built observatories, our observational facilities, the BVRI photometric data that we acquired for the Orion project, and our future plan for observation of variable stars of interest.

A Report on West Mountain Observatory Observations for the KELT Follow-up Observing Network

Michael Joner
Brigham Young University, Department of Physics and Astronomy, N488 ESC, Provo, UT 84602; xxcygni@gmail.com

Abstract The Kilodegree Extremely Little Telescope (KELT) project is a ground-based observational system that is dedicated to searching for exoplanet planet candidates $8<V<10$ by photometrically detecting suspected transit events. Brigham Young University astronomers have participated as part of the KELT Follow-up Observing Network for the past year with observations from various small telescopes including the three small research telescopes located at the West Mountain Observatory. This presentation will report on the structure of the KELT project and examine some of the observations that have been made with telescopes at the West Mountain Observatory.

Visual Observing: New Ideas for an Old Art?

David Turner
St. Mary's University, Department of Astronomy and Physics, 56 Shalimar Crescent, Dartmouth, NS B2W 4L8, Canada; turner@ap.smu.ca

Abstract New detectors have had a positive effect on the precision of observations by amateur observers, but often overlooked is the fact that new methodologies can also improve the precision of simple eye estimates. Described are a variety of techniques used successfully to make visual observing easier and more reliable than is sometimes the case. There are many variable stars for which such techniques could greatly improve the scientific value of the observations for astronomical analysis. Some are often too bright for standard photometric techniques.

America's First Variable Star

John Toone
17 Ashdale Road, Cressage, Shrewsbury, 5Y5 6DT, England; enootnhoj@btinternet.com

Abstract An account of the mistakes, controversy, and confusion associated with T Leo (QZ Vir), the first variable star to be discovered from the USA.

General Paper Session Part IV

The Future of Visual Observations in Variable Star Research: 2015 and Beyond

Mike Simonsen
AAVSO Headquarters, 49 Bay State Road, Cambridge, MA 02138; mikesimonsen@aavso.org

Abstract In this paper we examine the strengths and weaknesses of visual observations in variable star research and outline areas where visual observers can still make a contribution to science. We also examine reasons for continuing to support visual observers' participation in the AAVSO for decades to come.

The Life of Albert Jones

John Toone
17 Ashdale Road, Cressage, Shrewsbury, 5Y5 6DT, England; enootnhoj@btinternet.com

Abstract A biographical account is given of Albert F. A. L. Jones, the world's most prolific visual variable star observer.

Special Paper Session Part I

Why Do Some Cataclysmic Variables Turn Off?

Kent Honeycutt
Indiana University, Astronomy Department, 727 E. 3rd Street, Bloomington, IN 47405; honey@astro.indiana.edu

Abstract VY Scl Stars are a class of CVs which fade by as much as five magnitudes at irregular intervals. Dedicated long-term monitoring programs are needed to study such phenomena. We discuss VY Scl stars and other kinds of long-term CV variability, with an emphasis on collaborations with Arne Henden in these studies. We also discuss Arne's work with early low-light-level detectors, experience that has contributed to his broad knowledge of photometric techniques that has been so valuable to the AAVSO. We then show observational evidence that starspots on the secondary star are responsible for CV low states.

Special Paper Session Part II

Before the Giants: APASS Support to Ambitious Ground-based Galaxy Investigations and Space Missions Serching for Exo-Earths

Ulisse Munari
INAF—Astronomical Observatory of Padova, Vicolo dell'Osservatorio 5, Padova I - 35122, Italy

Abstract A huge, worldwide effort is underway to reconstruct the structure, kinematics and evolution of our Galaxy with optical spectroscopic techniques, which provide radial velocities and individual chemical abundances in addition to deriving fundamental stellar parameters like surface temperature and gravity. For ten years (2003–2013) the Radial Velocity Experiment (RAVE) has used the 6dF 150-fiber positioner at the 1.3-meter UK Schmidt telescope in Siding Spring. ESO-Gaia has hundreds of nights allocated at the VLT telescopes in Chile with UVES and GIRAFFE multi-fiber instruments. The Galactic Archaeology with HERMES (GALAH) survey has been allocated 400 nights in five years with the 400-fiber High Efficiency and Resolution Multi-Element Spectrograph (HERMES) at the 4-meter Anglo-Australian Telescope. Common to the millions of stars targeted by these surveys (over the range $10 < V < 16$ mag) is the lack of suitable, multi-band, accurate optical photometry. In this talk, I review the fundamental role played by the AAVSO Photometric All-Sky Survey (APASS) in providing such missing photometric information for the stars targeted by these gigantic spectroscopic surveys. The APASS BVgri data are fundamental to support the spectroscopic effort, for example to constrain (when modelled together with 2MASS infrared JHK photometry) the stellar temperature. The APASS data are also crucial in fixing the interstellar reddening and the distance to the target stars, and their importance will be further expanded when APASS ultraviolet (u) and far red (z,Y) magnitudes will become available, as well the unsaturated APASS extension to brighter stars so that most of the bright spectroscopic standards will become within photometric reach.

APASS and Galactic Structure

Stephen Levine
900 E. Hilltop Avenue, Flagstaff, AZ 86001; selevi@gmail.com

Abstract While the AAVSO Photometric All-Sky Survey (APASS) catalog was designed to facilitate all-sky photometric calibration, especially for variable star work, it can be used for much more. This will be a short look at some of the aspects of local galactic neighborhood kinematics and structure that can be studied using APASS data.

Astronomical Photometry and the Legacy of Arne Henden

Michael Joner
Brigham Young University, Department of Physics and Astronomy, N488 ESC, Provo, UT 84602; xxcygni@gmail.com

Abstract Arne Henden has helped provide a valuable resource to the photometric community with the publication of the 1982 book *Astronomical Photometry*. I will present a brief review of the topics covered in this handbook and recount some of the many times that it has been useful to myself and my students for answering a wide variety of questions dealing with the acquisition and reduction of photometric observations.

Special Paper Session Part III

A Journey through CCD Astronomical Imaging Time

Richard Berry
22614 N. Santiam Highway, Lyons, OR 97358; rberry@wvi.com

Abstract The author describes his experience in astronomy and as a pioneer in CCD imaging and astronomical image processing, and discusses Arne Henden's contributions to the field.

Collaborations with Arne on Cataclysmic Variables

Paula Szkody
Department of Astronomy, University of Washington, Box 351580, Seattle, WA 98195; szkody@astro.washington.edu

Abstract The start of the Sloan Digital Sky Survey in 2002 marked the beginning of a 14-year-long collaboration with Arne Henden on the photometry of cataclysmic variables. Starting with the USNO Flagstaff station, and continuing with AAVSOnet, Arne and the AAVSO members contributed ground-based followup of SDSS candidate CVs to determine their orbital periods and characteristics. In addition, many scientific studies using spacecraft observations with HST, XMM, and GALEX were enabled and improved due to their contemporaneous ground-based photometry. Some of the primary results in the 39 publications resulting from this long term collaboration will be summarized.

The History of AAVSO Charts, Part III: The Henden Era

Mike Simonsen
AAVSO Headquarters, 49 Bay State Road, Cambridge, MA 02138; mikesimonsen@aavso.org

Abstract In this paper we pick up where "The History of AAVSO Charts, Part II: The 1960s Through 2006" (Malatesta *et al.*, 2007, *JAAVSO*, 35, 2, 377) left off and discuss the evolution of the automated chart plotter, the comparison star database, the new tools available to the chart and sequences team, and Director Arne Henden's influence and legacy.

Arne's Decade

Gary Walker
21 Ashley Lane, P. O. Box 26, North Groton, NH 03266; bailyhill14@gmail.com

Abstract An affectionate and humorous look back is given on Arne Henden's ten years as AAVSO Director.

NOTES

Made in the USA
Middletown, DE
04 September 2023